Luchesi Lincoln &

Fine Surgical and Orthopaedical Instruments

Luchesi Lincoln &

Fine Surgical and Orthopaedical Instruments

ISBN/EAN: 9783337172152

Printed in Europe, USA, Canada, Australia, Japan

Cover: Foto ©berggeist007 / pixelio.de

More available books at **www.hansebooks.com**

NOTICE.

.... With this we hand you our ....

REVISED

ILLUSTRATED PRICE LIST

which we hope will serve you as a reference.

The prices in this catalogue are **NET**.

Goods ordered by parties unknown to us
will be sent C. O. D.

Articles of our manufacture are all of the
best quality of material and workmanship.

We are continually adding new instruments
to our list, and will from time to time send you
circulars of them.

Very respectfully yours

Lincoln & Luchesi,

143 EAST 23D ST.,

NEW YORK.

FORMERLY WITH

J. REYNDERS & CO.

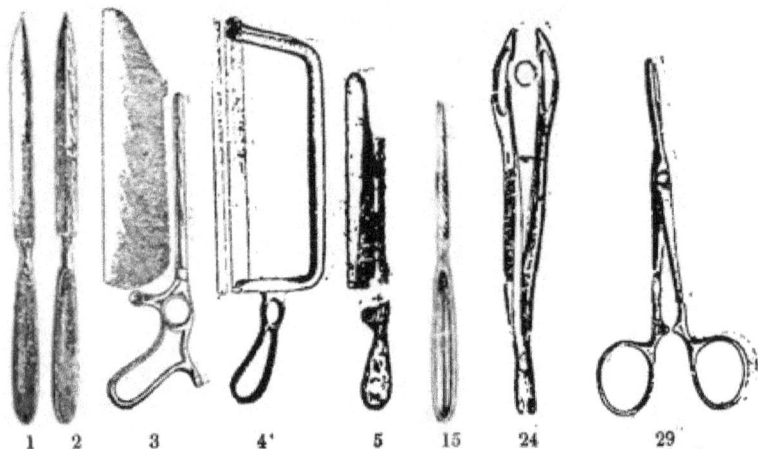

1   2   3   4'   5   15   24   29

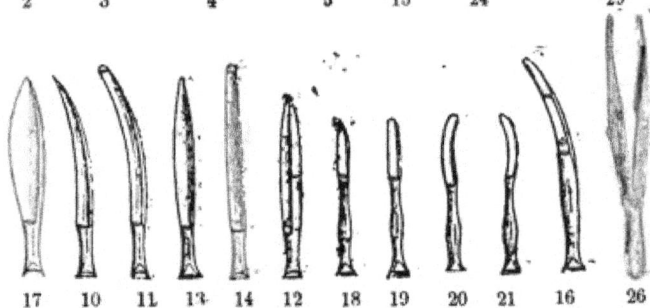

17   10   11   13   14   12   18   19   20   21   16   26

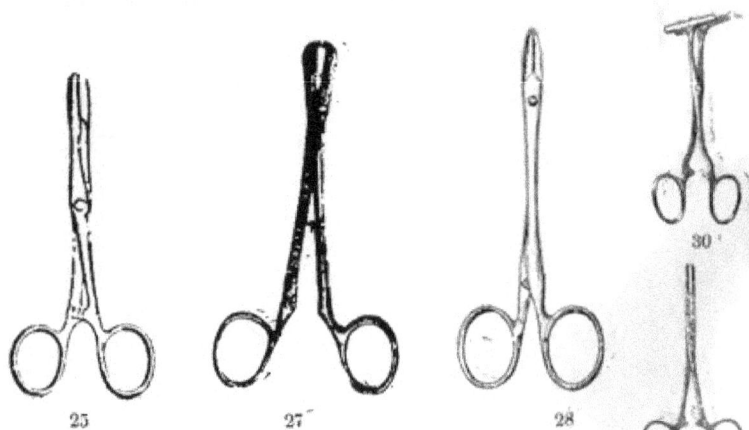

25   27   28   30   33

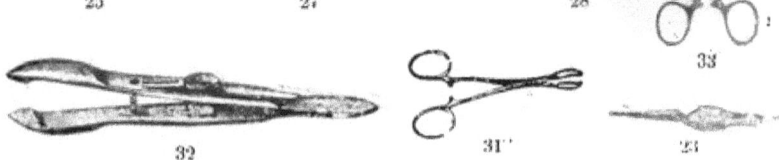

32   31   23

4'

Amputating Instruments.

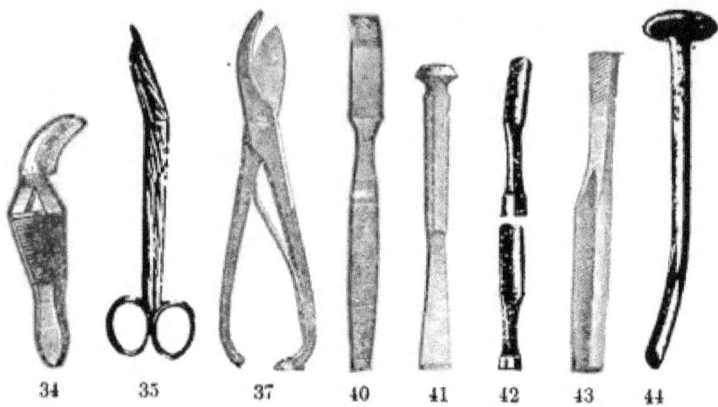

34   35   37   40   41 42   43   44

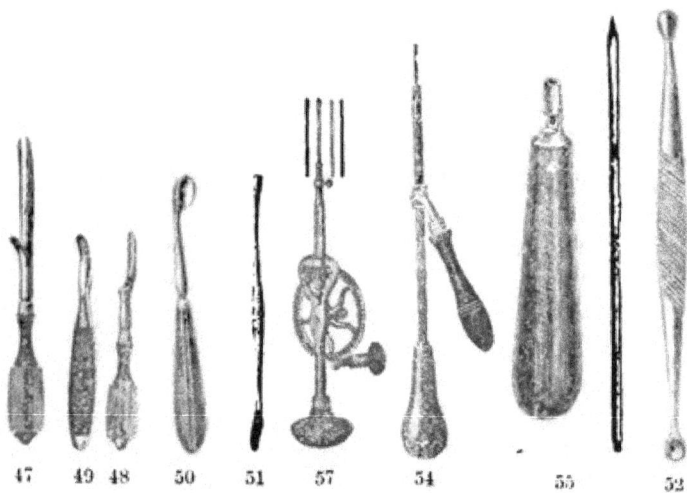

47   49 48   50   51   57   54   55   52

64   69

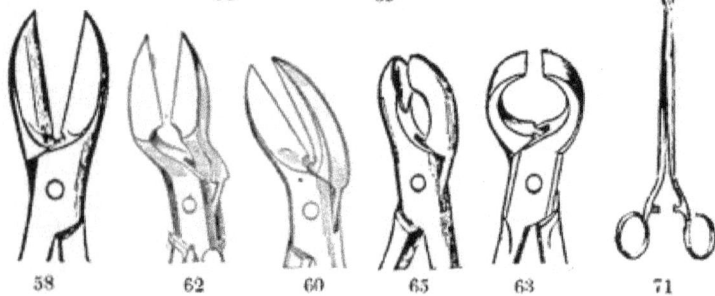

58   62   60   65   63   71

Artery Forceps, plain 50
" " " with spring-catch 75
23 " " Allis straight or angular 1 00
" " fenestrated with spring-catch ... 1 00
" " " " " bull-dog 1 25
" " " " slide 1 50
24 " " fenstr., with needle-holder at other
end, slide-catch 1 75
" · " bull - dog fenestr., with needle-
holder at other end 1 75
25 " " Becks 1 50
26 " " Frickes 1 50
27 " " Phelps 1 50
28 " " Lawson Taits ..... 75 per doz. 8 00
29 " " Woods or Peans ... 75 " 8 00
30 " " Peans T. 2 00
31 " " " fenest 2 00
" " Wylie's curved ....$1 25 per doz.13 00
32 " " Esmarchs 1 50
33 " " Wyeths 1 50
34 " Clamps, self-closing straight 35
" " " curved 35
35 Bandage Shears small 1 50
" " medium 1 75
" " large 2 25
" " Phelps 3 50
37 " " Suetains 3 00

Operating Instruments.

40 Bone Chisel 6 inch 65
" " 8 " 85
41 " " MacEwins 1 50
42 " Gouge " 1 75
43 " " 6 inch 70
" " 8 " 85

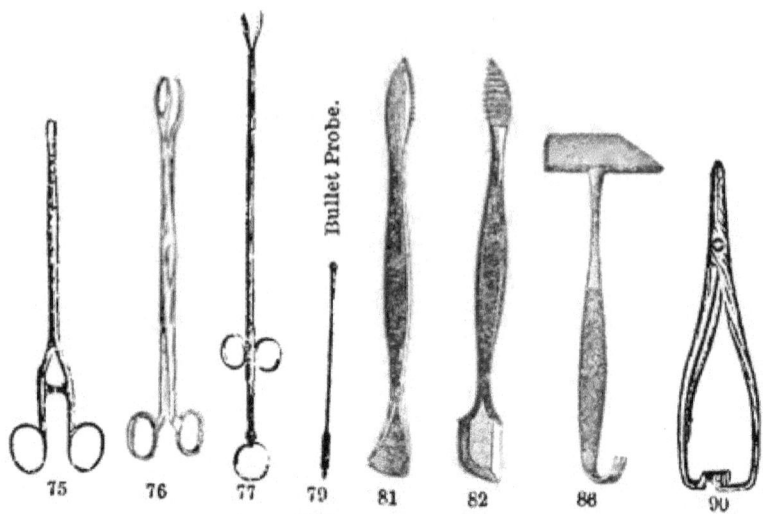

Bullet Probe.

75   76   77   79   81   82   88   90

97

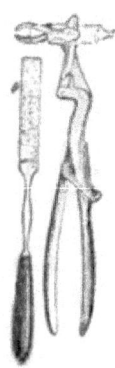

94   96   98   102   100   99   101   103

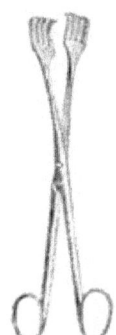

105   108   110   8   115   111

9·

Fig.

86 Mallets, Steel 1 00
    Needles, straight, half or full curved, per dozen .. 50
        "   Hagedorns, per doz. 75
        "   aneurism, Mott's or Student's 1 25
        "   in handle, light or heavy 75
    Needle-holder, Sims 1 50
90   "   "   Hanks 2 00
        "   "   Lewis 2 00
        "   "   McBurneys 2 00
        "   "   Reiners 2 25
94   "   "   Russian 2 00
        "   "   Stimson 2 00
96   "   "   Skenes 4 00
97   "   "   Wyeths 3 00
98 Osteotome Wyeths 12 00
99 Retractors Billroths 1 25
        "   Buck per pair 75
        "   Bulls 1 25
100   "   Esmarchs each 1 00
101   "   Langenbeck's blunt 3 pronged 1 00
        "   "   " 4 " 1 00
102   "   McBurneys for skin Grafting 2 00
103   "   Mt. Sinai "Giant" 2 25
        "   "   "   major 2 75
105 Retractors, Parker's or Mott's, per pair 75
        "   Volkmann's 2 pronged, each 1 25
108   "   "   3   "   " 1 25
        "   "   4   "   " 1 25
        "   "   5   "   " 1 50
110 Retractor forceps 3 pronged 2 00
        "   "   4 or 5 pronged 2 50
111   "   "   Byrnes 3 00
    Wyeths skewer each 1 25
    Tourniquet Field 60
115   "   Petits 1 25

117

116

119

121

149

125

135

140

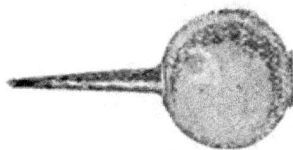

153

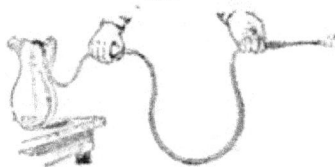

145

147

155

156

Fig.
116 Tourniquet Clamp Esmarchs 1 50
117 Tourniquet Esmarchs, complet e................. 2 25
119 Trephine Galts conic 2 00
" " aseptic 2 50
" Roberts 3 sizes....................... 3 50
121 Trocar and Canula, with spout 1 00
" " reversible.................. 1 25
" " 3 in set.................... 2 00
" Exploring 1 00
" " Bulb 50

Absorbent Cotton.

Size of Packages	1 lb ℔ lb	½ lb ℔ lb	¼ lb ℔ lb	2ozs. ℔ lb	1oz. ℔ lb	½oz. ℔ lb
Plain Absorbent..	$ 50	$ 60	$ 65	$ 70	$ 75	$ 75
Borated	60	75	75	80	85	85
Carbolated.......	60	75	75	80	85	85
CorrosiveSublim..	60	75	75	80	85	85
Salicylated	70	85	85	90	95	95

Absorbent Gauze in glass jars

	1 yd x 3 yds	1 yd x 5 yds	1 yd x 100 yds
Plain........	——	——	$ 6 50
Sublimatized,1 to 2000	$ 18	$ 65	7 50
Borated, 10 ℔ ct.	20	75	10 00
Carbolated, 10 ℔ ct.	18	65	8 00
Iodoform, 10 ℔ ct.	30	1 25	25 00
" 20 ℔ ct.	40	1 60	
" 25 ℔ ct.	50	2 00	
" 33½ ℔ ct.	60	2 50	

Hereafter all our Iodoform Gauze will be prepared by a
new and improved formula, which leaves the Gauze
dry; or nearly so; thereby increasing its value and
decreasing the risk of deterioration by age to the
minimum. We shall fill all orders in this way unless
"MOIST" is specified on the order.
125 Absorbent Gauze boxes, add to above prices......$0.25

Roller Bandages.—Unbleached Muslin.

per doz.		per doz.	
1 in. x 1 yd	10	3 in. x 4 yds	48
1 in. x 3 yds	18	3 in. x 8 yds	1 25
1½ in. x 3 yds	25	3½ in. x 5 yds	85
2 in. x 3 yds	34	3½ in. x 8 yds	1 50
2 in. x 8 yds	90	4 in. x 6 yds	1 25
2½ in. x 3 yds	40	4 in. x 8 yds	1 48
2½ in. x 8 yds	1 00		

Anticeptic Gauze Bandages.

per doz.		per doz.	
1 in x 3 yds	10	2½ in x 8 yds	90
2 in x 3 yds	30	3 in x 8 yds	1 20
2 in x 8 yds	70	4 in x 8 yds	1 50
2½ in x 3 yds	40		

Elastic Bandages.

2 in x 3 yds each	50	2½ in x 5 yds each	75
2 in x 5 yds "	65	3 in x 3 yds "	75
2½ in x 3 yds "	60	3 in x 5 yds "	1 00

Rubber Bandages with Tapes.

2 in x 3 yds	50	3 in x 3 yds	75
2½ in x 3 yds	65	3 in x 4½ yds	1 00
2½ in x 4½ yds	90	3 in x 6 yds	1 25

135 Plaster of Paris Bandages, 3 in x 4 yds each 15

3 in x 6 yds " .. . 22

3 in x 8 yds " 25

Bandage Roller wood 1 25

" " Jobse 3 00

140 " " Patent 3 00

Catgut dry Nos. 00—0—1—2—3 or 4 per coil 06

" " per doz. ....

" 3 spools in bottle put up in Juniper Oil,

Sublimated, Carbolized or Chromatizd each 50

Catgut Hospital size per bottle 1 50

Drainage tubes glass each 12

13

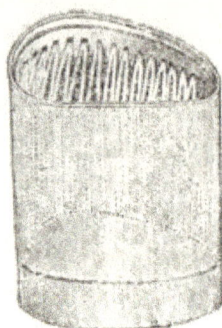

160   165   166

170   168   180 181

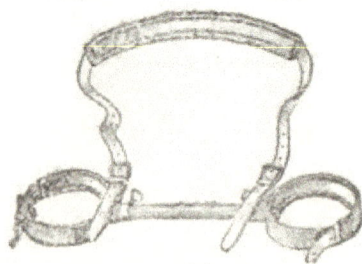

191

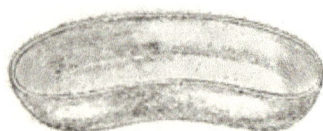
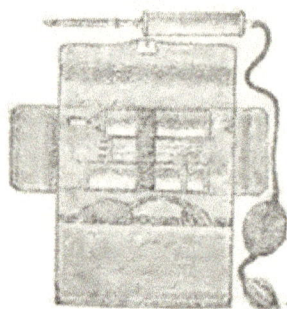

190

200   205

208

SURGEONS'
Pure Silk, Iron Dyed.
No. 11.

211

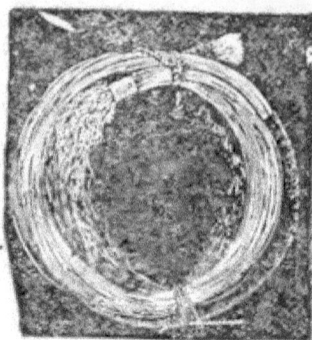

215

220

236

THREE SIZES ON A TABLET

212

232

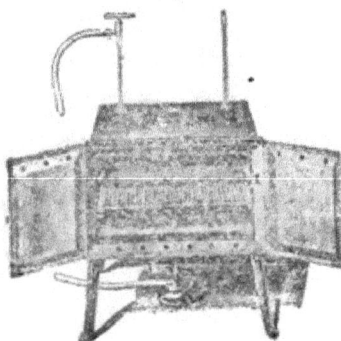

237

238

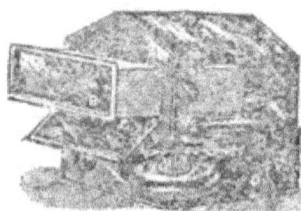

240

each

									each
Pus Basin, Papier Mache, finely Japanned, 16 ozs   50

"   "   "   "   "   "   24 "   50

"   "   "   "   "   "   34 "   60

"   "   "   "   "   "   50 "   75

"   "   "   " Triangular "   34 "   1 00

"   "   "   " Fan Shaped "   50 "   00

"   "   Hard Rubber, 16 ozs................. 1 00

"   "   "   "   24 " 1 25

"   "   "   "   34 " 1 50

Paquelins Cautery with two Platinum tips in
moracco case 17 00

205 Paquelins Cautery in pocket case form complete..17 00

Paquelins latest style Cautery, so arranged that the
Points cannot burn out or become clogged up ..25 00

Circular on Application.

208 Rubber Apron................................ 2 50

"   " Kelleys 3 50

211 **Silk:**—Reels, black or white, twisted, 14 sizes,
each, 0.10, per doz 90

Tablets, in slip case, same, one size, twisted, 0.20,
per doz. 2.00

212 Tablets, in slip case, 3 sizes 30

Turner's English Braided Silk, 4 sizes, in slip tablet   50

Pedicle Silk, chemically pure, the strongest, per
pool.................................... 50

English Braided Silk, in hanks, Nos. 1 to 12,
from 0.10, to 50

In Antiseptic Solution, 3 sizes, in screw top bottle   50

Same, hospital size........................ 1 25

215 Silk worms gut per 100 Strands.............. 75

Silk protective amber, per yd. 1 00

"   " green, " " 1 25

220 Silver wire per 1 yd. spool size 24 to 32 each..... 20

"   " per oz............................. 3 00

Fig.

```
Sponges small, per doz. .......................  4(
   "   No. 2  "   "   .......................  6(
   "   flat No. 1 each ....................  2£
   "    "    " 2  "   ......................  5(
   "    "    " 3  "   ......................  7£
   "    "    " 4  "   ......................  9(
   "   1 doz. in bottle No. 1 .................  1 0(
   "   1  "   "    "    " 2 .................  7£
```
230 Sponge jars w. covers:

```
      3 inch x  7....$0 50 ....4½ x  8 ......  7£
      3 inch x 10....  60 ....4½ x 11½ ......  8(
      4½ inch x  6....  65 ....6  x  8 ......  1 3£
                              6  x 12 ......  1 5(
```
232 Sponge dish, 9 inches .... 1 2£

Sterilizer Milk:—Family size, Arnold's No. 1, with

   7 Bottles, ... 3 0(

   ,,   Arnold's General Purposes.—Sizes of Sterili
zing Chambers and Prices—

	Height.	Diameter.	Heavy Tin, Dopper Bottom.	All Copper
No. 2..	7¼ ins.	8½ ins.	$2 50	$7 0(
No. 3..	10¼ "	9⅜ "	3 00	7 7£
No. 4..	11½ "	10¼ "	3 50	8 7£
No. 5..	12½ "	11¼ "	4 00	9 5(

Sterilizer Arnold's Special for Instruments:—Oval in
Shape; size of Chamber, 14 ins. long, 9½ ins. wide, (
ins. deep, with two Racks for instruments. Heavy tin
copper bottom. $5.00; all copper, $12.00.

237 Sterilizing Oven:—Made for Sterilizing Instruments
Bandages, etc., in Dry Heat. It has been our aim
to construct an oven to produce the most uniform
heat inside of the oven. They are made of the bes
Russian iron, with double walls, through which th
heat circulates, having a register on the top. In
order to secure a perfectly even circulation of th
heat all around there are connecting openings betweei

18

the double doors. The heat is furnished by a powerful Bunsen Burner, which is placed below the tunnel-shaped opening in the center of the bottom of Oven, circulating through the double walls and escaping through the register in the center of the top. In order to secure the proper heat necessary for Sterilization, and to prevent the overheating of the instruments, etc., which would spoil them, there is inserted into the opening on the one side in the top of the Oven, and held by a cork, as shown in the cut, an improved Gas Regulator. These Ovens are made in four sizes:

Inside measure 9x	9x12	incl.BunsenBurner,Gas	$16 00
" "	9x 9x15	Regulator, Thermomet.	20 00
" "	9x 9x18		23 00
" "	12x12x24	" "	30 00

Improved Gas Regulator........................ 3 00

Thermometer, 400° F......................... 1 50

238 Dr. Rotter's Compact Sterilizing Apparatus:—For Boiling Instruments in 1 per cent. Soda Solution. Consists of 3 Dishes, with Cover, of the following dimensions: No. 1 Dish, 6½ ins. long, 4½ ins. wide, 1¼ ins. deep; No. 2 Dish, 9 ins. long, 5 ins, wide, 1½ ins. deep; No 3 Dish, 13 ins. long, 6 ins. wide, 2¼ ins. deep (each Dish with Cover and inside Tray for lifting out instruments); 1 Folding Support on 4 legs: 1 Alcohol Lamp of peculiar construction, rendering sufficient heat to sterilize instruments within 4 minutes. The apparatus replaces by its compactness and cheapness any other Sterilizer in the market. It is made of sheet iron, heavily tin lined. Price..........$10 00

240 Van Heusen's Sterilizer. No. 1, 7½ ins. by 11½ ins., full copper, nickel trimmings, with forced flame lamp...................................... 8 00

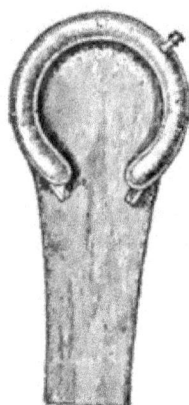

260

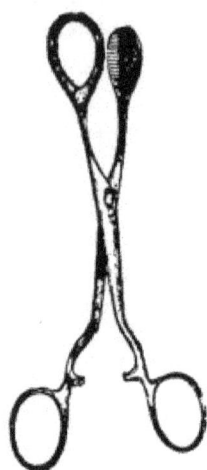

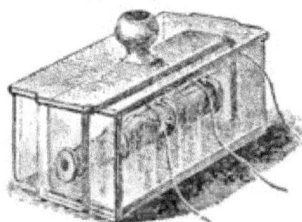

245

265

2(9

256

286

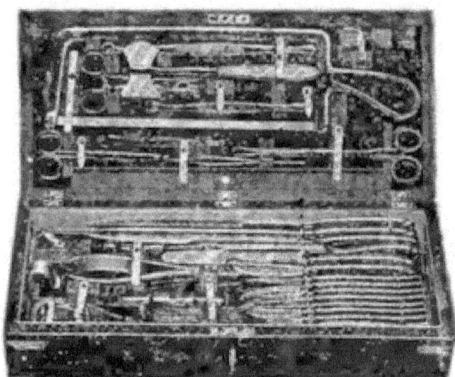

365

20

No. 2, 8 ins. by 16 ins., full copper, controllable new triple forced flame lamp, nickel trimmings.. 14 00

No. 3, 12 ins. by 24 ins., full copper, controllable triple forced flame lamp, nickel trimmings.... 25 00

Send for Circular.

245 Surgical Cushions, Kelley's small			3 50
" " " med. 18 in.			4 50
" " " large 21 in.			5 50
" " Chapmans			6 00
Suture Pins, glass heads, per doz			35
" " Carlsbad, per 100 ass.............			40
" " lance point, per 100			1 00
" " Needles, straight, half or full curved per doz.....................................			50
Suture Needle Hagedorn per doz			75
260 Suture box H. R.			1 50
" " glass, 3 spools....................			1 50
" " " " "			3 00
265 " " Hospital Size......................			5 25
" " 4 spools, Hospital Size			7 00
269 Tongue seizing forceps. Esmarchs..............			1 50
" " " Houz...................			3 00
276 Trays, glass, 6 x 8 0.75, 8 x 10 1.00, 12 x 18			3 50
" agate, 7½ x 9 0.60, 8 x 11...............			80
" " 9 x 13 0.90, 14 x 18.............			2 00
" sof rubber, small 2.50 med.			3 00

Amputating, General and Minor Operating Cases.

The following Sets have ALL METAL HANDLES, as also PIN or FRENCH LOCKS so that the Instruments can be taken apart for boiling indefinitely without damage. The HARDWOOD CASES are polished on out and inside, or leather lined inside as desired.

Fig.

The Morocco cases are made of fine leather and velvet lined.

286 Amputating Case No. 1, contains: Listons long Knife; Listons medium Knife; Catling, Scalpel; Tenaculum; Taits Artery Forceps; Metacarpal Saw; L. & L. Saw; Tourniquet; Bone Forceps and one dozen Needles; Wax, Pins & Silk; in a black walnut case. Price $18 00

Amputating case No. 2 contains Liston's long knife; Liston's medium knife; Catlin; Satterlee's Amputating Saw; Metacarpal Saw; Scalpel; Tenaculum; 2 Taits Artery Forceps; Petits Tourniquet; Hair-Lip and Suture Pins; Needles and silk; in a mahogany case $20 00

Amputating and Trephining case No. 3: contains ListonsKnife long;Listons Knife medium; Catlin, Metacarpal Saw; Capital Saw; Heyes Saw; Galts Trephine; Petits Tourniquet; Elevator and Raspatory; Scalpel; Tenaculum; 2 Taits Artery Forceps: Listons Bone Forceps; Plastic Pins; needles and silk; catgut in a mahogany. Case Price.....................$27 00

289 Amputating Case No. 4 contains, 1 Amputating Knife, Leg and Arm, 1 Finger Knife, 1 Hernia Knife, 1 Sharp Curved Bistoury,2 Scalpels, 1 Tenotome, 1 Tenaculum, 1 Pair Scissors, curved, 1 Finger Saw, 1 Capital Saw, 1 Liston's Bone Forceps, 1 Artery and Needle Forceps combined, 1 Dressing Forceps, 1 Esmarchs Tourniquet with chain, 1 Director, with Aneurism Needle, 1 Pair Probes, 1 Tablet Silk, four sizes, in slip case, 1 Coil Silver Wire, one-half dozen Assorted Needles in a Mahogany Case polished in and outside, Knives fitted in a METAL-RACK.....................$22 00

The above with Aseptic Trephine and Trephine, Elevator 24 75

The above with 6 Taits Artery forceps......... 28 75

Operating case No 5 Parkers, contains Liston's long Knife, Listons small knife, Catlin, Metacarpal Saw,

Capital Saw, **Heyes'** Saw, Scalpel large, Scalpel medium, Probe pointed curved Bistoury, Finger knife, Tenaculum, Taits **Artery** Forceps, Elevator and Raspatory, **Trephine,** Tourniquet, Bullet Forceps, Listons Bone **Forceps,** Director, Aneurism Needle, Bullet Probe, **Dressing** Forceps, Straight Scissors, one pair Silver Probes, Plastic Pins, Needles, Silk, Catgut, in a mahogany Case, price.....................$36 00

With 6 Taits **Artery** Forceps, 1 Volkmanns double bone scoop..$40 00

General Operating Case No. 6. Jas. R. Woods, contains: long Ampt. knife; circular **knife;** Catling; large **Saw;** Tenaculum; Taits Artery Forceps; Liston's Bone Forceps; **Screw Tourniquet;** American **Bullet** Forceps; Galt's **Trephine** and handle; Trephining Scalpel and Raspatory; Trephining Elevator; Heys' **Saw;** Brush; Finger Saw; Straight Dressing **Scissors;** curved probe point **Bistoury;** curved sharppt. Bistoury; 2 **Scalpels;** Aneurism Needle; steel **Director;** pair Silver **Probes;** curved Eye **Scissors;** Beer's Cataract **Knife;** Strabismus Hook; curved Strabismus Forceps; straight Eye **Needle;** curved Eye **Needle;** 2 steel Sounds; silver-plated Catheter; Double Web Catheter; 2 **Elastic Bougies;** Needles; Silk; etc. In a rosewood, brassbound case, polished in and outside Price..... $42 50

With 6 Taits forceps 1 Volkmanns double bone scoop. 1 Bullet probe..............................$47 50

305 Operating Case No. 7; contain Bow **Saw;** Medium Amputating **Knife;** Medium **Catling;** Finger **Knife;** Metacarpal **Saw;** Hey's **Saw;** Amputating **Scalpel;** Amputating **Tenaculum;** Straight Bistoury; 2 Curved Bistouries; Tenotome; 2 Scalpels; Aneurism Needle; **Coopers Hernia Knife;** pair Retractors; Bone **Forceps; Sequestrum Forceps;** Pin-cutting Forceps; 6 Taits Artery Forceps; Hanks Needle Holder; Vulsellum

Forceps; Spring Forceps; Straight Scissors; Curved Scissors; Volkmanns Double Scoop; Petits Tourniquet; Trephine, Brush; Elevator; Trocar; Director; Bullet Probe; 2 Probes; Needle; Silk; Catgut and Silver Wire. in a brass-bound; polished in and outside rosewood case; Knives in a Metal movable rack ...$56 00

General Operating Case No. 8: Hamiltons contains; Catling large; Catling medium; Liston's Knife for arm and leg; large Saw; slide-catch Artery and Torsion Forceps; Hamiltons Artery Forceps; Tenaculum; Listons Bone Forceps; pair Motts Retractors; American Bullet Forceps, Bullet Probe; Screw Tourniquet; Field Tourniquet; 3 Acupressure Needles; 6 Serrefines; narrow Saw; probept. curved Bistoury; broad French Bistoury; narrow French Bistoury; straight Dressing Scissors; Dressing Polypus Forceps; Director and Ear Scoop; set Motts Aneurism Needles; plain Artery Forceps; extra long plain Artery Forceps; Conic Trephine; Necrosis Trephine; Handle for both; Trephining Elevator and Raspatory; Heys Saw; Tirefond; Steel Sound; 2 male Catheters; Harelip Scissors; h. r. Dental Syringe; Saw Brush; Needles; Silk, etc. In a fine rosewood brass-bound case, polished in and outside Knives in a metalrack and patent leather cover $60 00

With 6 Taits Artery Forceps; 1 Volkmanns Bone Scoop. Price $65 00

Operating Bag for Antiseptic Surgery; Gersters containing:

1 Bottle, 2 ozs. in wood box, with Corrosive Sublimate. box marked: "Corrosive Sublimate, 1: Alcohol, 10; 1 teaspoonful to a quart of water; strength, 1: 1500."

1 Bottle, 4 ozs. in wood box, Carbolic Acid, pure; marked "4 teaspoonsfull to a quart of hot water; shake well; strength, 3 per cent."

1 Bottle, 2 ozs., in corrugated tin box, Chloroform.

289

310

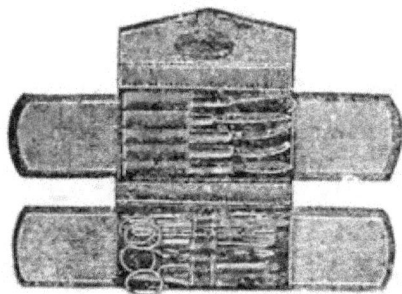

317

315

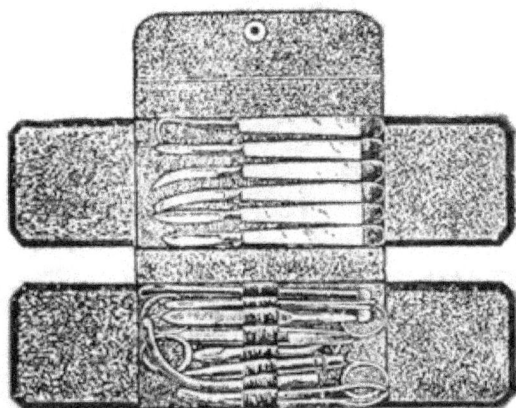

318

324

323

326

1 Bottle, 4 ozs., for Catgut Ligature, in corrugated tin box, with glass reels.

1 Hard rubber, Iodoform Spinkler.

1 Tin Can, with ½ ℔ of Ether.

1 Glass Jar, with rubber cork and metal screw top, for Rubber Drainage Tubing, with 5 parts carbolic solution to 100 parts water.

1 Glass Jar, for Compressed Carbolized Sponges.

1 Scissors, for cutting Gauze.

1 Tongue seizing forceps

1 Razor. 1 Teaspoon. 1 Piece of Wax.

1 Ether Inhaler and 1 Esmarch's Chloroform Mask, both enclosed in rubber cloth bag.

1 Fountain Syringe, with 3 glass points and rubber tubing; also in rubber cloth bag.

1 Hypodermic Syringe in metal case.

1 Stiff Nail brush, solid wood back.

1 Dozen large and small Safety Pins.

1 White Linen Pouch, for the reception of instruments.

1 Nest 6 Tin cups.

1 Nest 3 large Square Tin Basins.

The Valise has loops inside for holding the bottles and the various articles; an inside pocket for small articles and an outside pocket for dressings. The Valise is placed in the nest of tin basins, and is fastened to the same by two leather straps. Size of valise, 18 ins. long by 8½ ins. wide. Price; complete.......$45 00

310 Operating Aseptic Hand Bag, Buchanan's. The Antiseptic Satchel here shown is one which has been in use in general practice for several years, and we have no hesitation in saying that it contains all the appliances necessary for the antiseptic treatment of a capital, or accidental operation. It has also the advantages of lightness and portabillity. It contains the following: 2 one-ounce jars Carbolic Acid (cry-

stals); 1 bottle Corrosive Sublimate Tablets; 5 yards
Sublimated Gauze in tin case; 6 Roller Bandages (ass.)
1 Patented Package assorted Catgut on reels; 1 Patented Package assorted Silk on reels; 1 Tin Flask for
Anæsthetics; 6 assorted Rubber Drainage Tubes in
vial; 1 oz. Iodoform in Hard Rubber Sprinkler; 1
Razor; 1 Nail Brush; 1 Cake Soap in case; 1 Imp. Hard
Rubber Irrigating Apparatus; 1 Roll 1-inch Rubber
Plaster (10 yards); 2 ounces Absorbent Cotton, all in
Leather Bag (16-inch) Frame. Price complete..$12 00

Minor Operating Cases.

Minor Operating case No. 1 contains Scalpel large, Scalpel medium· Scalpel small, Sharp point curved Bistoury, Blunt point curved Bistoury, Finger knife, Hernia knife, Tenatome, Tenaculum, Artery and Needle
Forceps, Aneurism Needle and Director, Straight Scissors, Dressing Forceps, Plastic Pins, Needles, Silk, in
mahogany case polished inside and outside; metal
handled Instruments Price.................$12 50
Minor Operating case No. 2 contains Scalpel large, Scalpel medium, Scalpel small, Bistoury curved sharp
point, Bistoury curved blunt point, Finger Knife,
Hernia Knife, Tenatome, Tenaculum, Thumb Forceps,
Slide catch Artery Forceps and Needle Holder, Polypus Dressing Forceps Straight Scissors, Aneurism
Needle and Director, Caustic Holder, Male and Female catheter, Silver Probes one pair, Plastic Pins.
Needles, Silk, Catgut, metal handle instruments, case
polished in and outside price................$15 00
Minor Operating case No. 3 Gross's contains Scalpel
large, Scalpel medium, Finger Knife Hernia Knife,
Bistoury sharp point curved, Bistoury blunt point
curved, Tenatome sharp pointed, Aneurism Needle,
Finger Saw, Elevator and Raspatory, one pair Retrac-

tors Liston's Bone Forceps, Sequestrum Forceps Polypus Dressing Forceps, Vulsellum Forceps, one pair straight Scissors, one pair curved Scissors, Fenestrated Artery Forceps, Lithotomy Bistoury, Hernia Director, Tongue Tie and Director, Needles, Silk, Catgut, metal handle instruments, in a mahogany case polished inside and outside............................$25 00

With 3 Taits Artery Forceps.................... 28 50

or Operating Case No. 4, Jas. L. Little's, contains: 3 ass Scalpels, curved Bistoury sharp point, curved Bistoury probe point, Tenaculum, Hernia Knife, Thumb Forceps, Phelps' Artery and Torsion Forceps. Artery Needle, 2 Retractors, Trocar and Canula Scissors angular, Director, 2 silver Probes, Rongeur curved small size, Needles, Silk, etc., in a fine rosewood case, polished in- and outside $23 00

Pocket Instrument Cases.

Pocket case No. 1 one fold, knives have Aluminium handles, Scalpel and Tenatome, curved sharp and probe pointed Bistouries, 1 pair Probes, Director and Tongue Tie, Dressing Forceps, straight Scissors, Silk, Needles............................... 6 00

315 Pocket case No. 2 Gross, two fold; knives have Aluminum handles with slide catches, consisting of Scalpel and Tenatome, curved sharp and probe point Bistouries, Gum Lancet and Tenaculum, Needle Holder and Artery Forceps combined, straight Scissors, Polypus Dressing Forceps, Parker's Male and Female Catheter with Caustic Holder, Exploring Needle Gross Ear Instrument, Thumb Forceps, 1 pair Probes, Director and Tongue Tie, Needles, Silk, in a Morocco case,..$10 00

317 Pocket Case No. 3 L. & L. aseptic knives, Scalpel, Sharp and probe point curved Bistouries, Tenatome, Tenaculum, Artery and Needle Forceps combined,

Thumb Forceps, French joint Scissors, Peans ar ery Forceps, Parkers combined Male and Female Catheter with Caustic Holder, Director and Tongue Tie, 1 pair Probes, Needles, Silk, in a Morocco case. .$ 8 50

318 Pocket Case No. 4 Littles Aseptic knives, contains Scalpel, Sharp and Probe point curved Bistouries, Tenatome, Tenaculum, Finger Knife, Artery Forceps and Needle Holder combined, Thumb Forceps, curved on the flat French joint Scissors, Peans Dressing Forceps, Parker's combined Male and Female Catheter with Caustic Holder, Gross Ear Instrument, Director and Tongue Tie, 1 pair Probes, Needles, Silk in a neat Morocco case $10 00

Pocket Case No. 5 with our patent Aseptic Handle, Scalpel, 1 each Sharp and probe curved Bistouries, Tenatome Finger knife, Tenaculum, Taits artery forceps, Artery and Needle forceps with slide catch, straight Scissors, Thumb forceps, Parkers combined Male and Female Catheter, porte caustic, probes, Director and Tongue Tie, Needles and Silk in a onefold Morocco Case $12 00

Pocket Case No. 6, 2 patent Handles, Scalpel, Finger knife, 2 Bistouries (curved, sharp and probe pointed), Tenatome, Gum Lancet, Tenaculum, Exploring Needle in Screw Case, Parkers Male and Female Catheter, Gross' Ear Instrument, Director and Tongue tie, Artery and Needle forceps with slide catch, Peans Artery forceps, Thumb forceps, Curved on flat Scissors, 1 pair Silver probes, Needles and Silk in a fine Calf ˃ kin or Morocco Case $15 00

Our patent Handles which are placed in the above two Cases, can be rendered perfectly Aseptic and are readily removed from the handle. See Fig. 326.

322 Pocket Case No. 7 contains: 6 Artery Clamps, straight Scissors, Pean's Forceps, straight Knife (sharp pointed)

straight Knife (blunt pointed), pair two Hook Retractors, Peaslee's Needle, Metacarpal Saw, Volkmann's Spoon, 2 pairs Thumb Forceps, Grooved Director and Aneurism Needle, 2 Probes, small metal box, with tight cover, containing 6 Needles, 2 Bottles in a metal frame one containing Catgut, the other Silk and one soft rubber Catheter, 1 Silver porte Caustic, in a metal case covered by a canvas pouch Price......... $15 00

Pocket Case Instruments.

Caustic Holder Hard Rubber......30cts.........$	40
" " " " with silver burner..	60
" " " " all silver..........	1 50
" " " " plated with platinum burner	3 00
Catheters, Combination, male and female 3 parts plated...............75 silver...............	1 75
Catheters, Combination, male and female 4 parts plated...... ......1 00 silver...............	1 90
323 Catheters, Parker's, with port-caustic......plated	1 25
silver	2 75
Catheters, Female ......plated. 30 silver.......	65
324 Director and Tongue Tie	30
" " " " silver................	75
" " Ear spoon, plated................	50
" " Aneurism needle plated...........	40
" " " " silver	75
328 Ear instrument for foreign bodies Gross' A or B...	50
Eye " double for foreign bodies	1 75
330 Exploring Needle in metal case................	50
" " " Alum. handle	75
" Trocar, silver.......................	1 00
333 Forceps, Thumb	50
" Splinter..............................	50
" Dressing	75

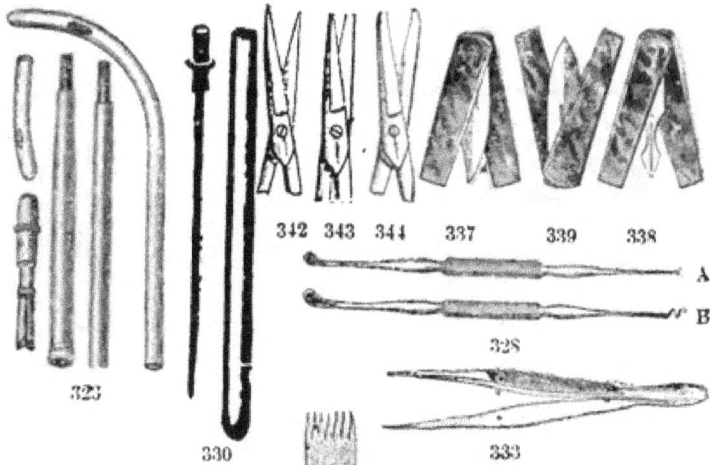

342 343 344 337 339 338

A
B

328

329

330

333

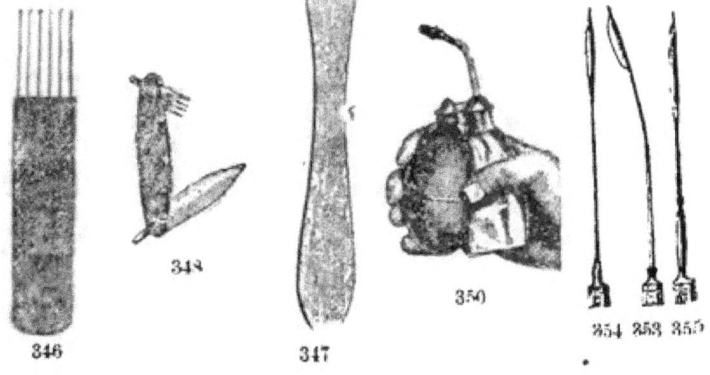

346 348 347 350 354 353 355

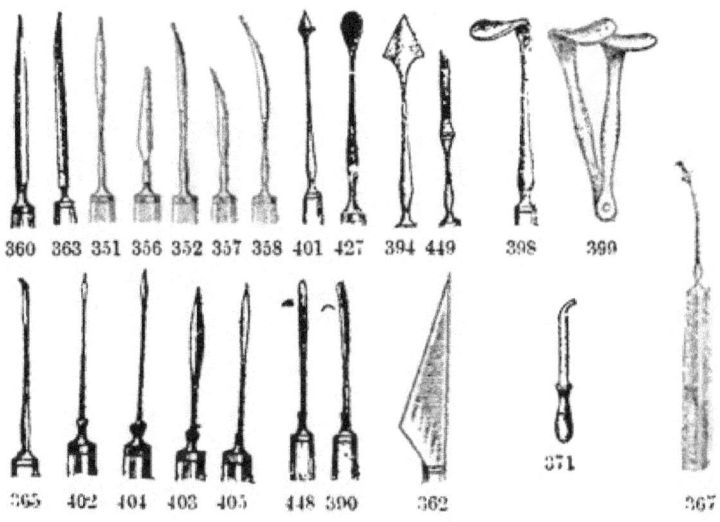

360 363 351 356 352 357 358 401 427 394 449 398 399

365 402 404 403 405 448 390 362 371 367

31

Forceps, (see **Amputating**).

Knives, single blades, folding in hard rubber handles 50
Knives, single blades, in shell handles 70
   "   "   "   " Alum. handles.. 75
   " double blades, in Aluminum or shell handles 1 50
Knives, double, in hard rubber handles 1 25
336   " blades patent aseptic, each 75
   " handle   "   "   " 1 00

(bracketed:) Knives any blade as per cuts (see amputating).

337 Lancet Abcess. h. r ....,50 shell or Alum....... 60
338   " spear pointed h. r ...50 shell or Alum 60
339   " Thumb h. r .... ..30 " " " 40

Needles (see Amputating)

Pocket Cases, empty, Morocco, 2-fold ..1 25 fine .. 1 50
   "   "   "   " 3 " ..1 50 " .. 2 00
   "   "   "   " 4 " ..2 00 " .. 2 50
   "   "   "   " Little's.1 50 " .. 2 00

Plaster Spatula, with Elevator or Scoop.... 50
   "   "   " Tongue-tie 50

Porte Mêche............................. 30
Probes, per pair ...... plated, 30 .... silver, 50
   " two parts, screwing together { plated.... 60
      { silver 1 00
   " Nelaton's, plated..................... 40
   " and Director, Hamilton's 1 00

342 Scissors, straight sharp points 4½ incl ....
343   "   " sharp and blunt 4½ incl 75
344   "   " blunt 4½ inch ....
   " as above 5½ inch 80
   "   "   " 6 inch...................... 1 00

Vaccine, fresh every other day, per box of ten points 1 00
Single points..... 15

346 Vaccine Comb............................. 20

347 Vaccine Comb Seibert's 25
343 " " and lancet, shell....50 Alum..... 60

Eye Instruments.

350 Atomizer, Agnew's 1 25
351 Canalicula Knife, Bowman's, probe pointed 1 00
352 " " " sharp pointed...... 1 00
353 " " Agnew's straight pliable shank.. 1 00
354 " " " probe pt. " " .. 1 00
      " " Jackson's..................... 1 25
355 " " Noyes probe pt. pliable shank... 1 00
356 " " Stilling's 1 00
357 " " Weber's short.................. 1 00
358 " " " long 1 00
360 Cataract Knife, Graefe's...................... 1 00
      " " Wecker's...................... 1 00
362 " " Beer's 1 00
363 " " Noyes 1 00
      Cystotome, Graefe's 1 00
365 " Knapp's 1 25
337 Cautery Iron Desmarres 1 25
      " Probe Gruenings.................... 1 50
      Director, Bowman's 50
      " Critchet's 50
370 Eye douche................................. 90
371 " Dropper 10
      " Shade single............................. 15
      " " double......... 25
374 Forceps, Cilia............................... 50
      " Iris straight 1 00
375 " half curved......................... 1 00
      " full " 1 00
      " Fixation 1 00
377 " " with catch 1 25
378 " Desmarres Entropium................. 2 00
      " Knapps Entropium right and left each.. 2 50

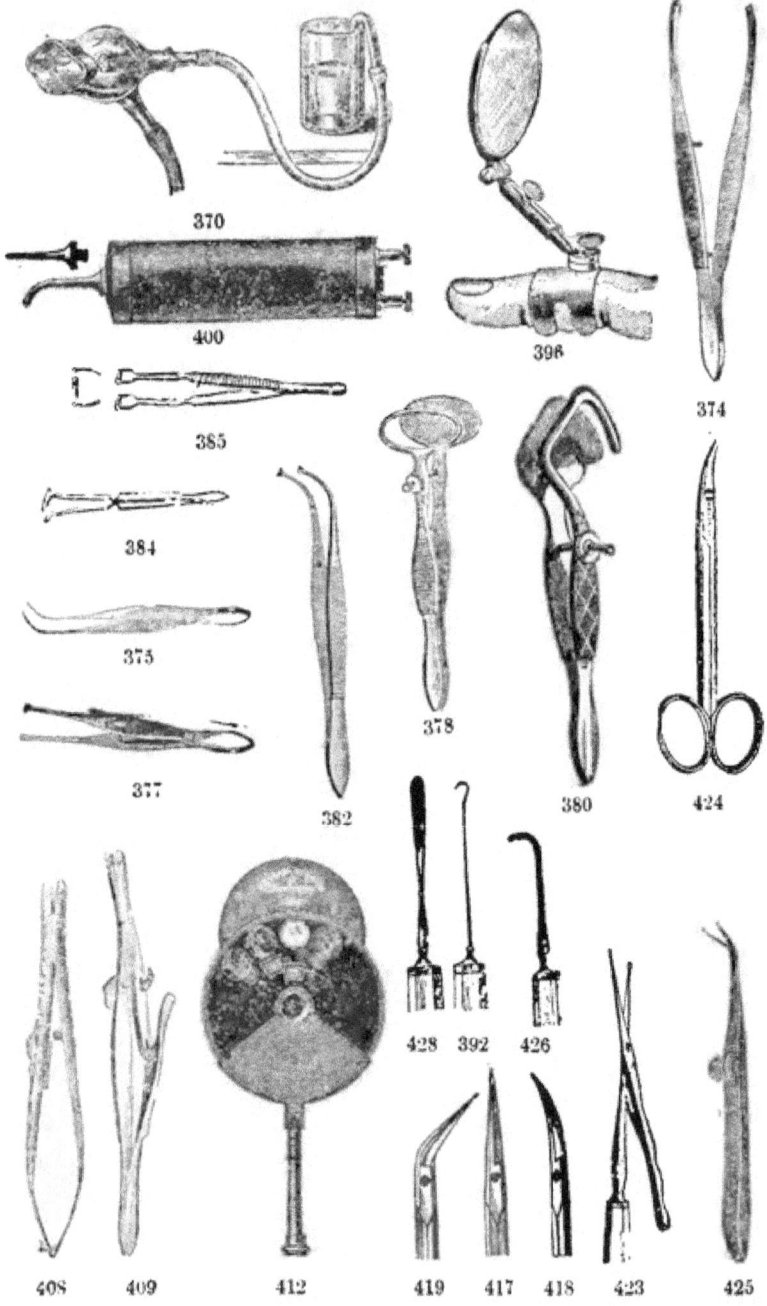

370

400

396

374

385

384

375

377

382

378

380

424

408 409 412 428 392 426 419 417 418 423 425

415

35

	Opthalmoscope, Knapp's 23 lens			15	00
	Probes Weber's				50
415	"	Bowman's Silver set, 8 sizes		2	00
	"	" Plated "		1	00
	"	Theobalds " 16 "		2	75
417	Scissors Iris straight			1	00
418	"	" curved on flat		1	00
419	"	" " angular		1	00
	"	Strabismus, curved		1	00
	"	" straight		1	00
	"	Enucleation		1	25
423	"	Noyes		3	25
424	"	Stevens		1	50
425	"	Wecker's		4	50
	Scalpels 3 sizes, each			1	00
426	Strabismus Hooks				75
427	Scoop Hard Rubber				75
428	"	Metal		1	00
431	Specula, plain				35
432	"	Graefe's		1	50
433	"	Liebreichs Aseptic		1	75
435	"	Mittendorfs			75
436	"	Noyes		1	25
437	"	Stevens		1	25
	Stevens Instruments.				
	"	Transfiction Needle		1	25
440	"	Tendon Hook		1	00
	"	Strabismus Hook		1	00
	"	Needle Holder		3	00
	"	Lid Retractor		1	25
444	"	Sub-conjunctival Tenotomy Scissors		1	50
445	"	Straight Iris Forceps		1	00
446	"	Curved " "		1	00
447	"	Straight " " with catch		1	75

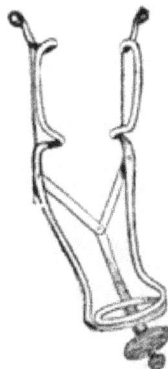
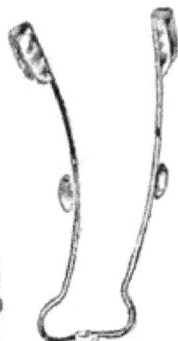

432   433   435   436   437

431   440   456   447 450 461 518 453   457

451

454   452   467

460

455
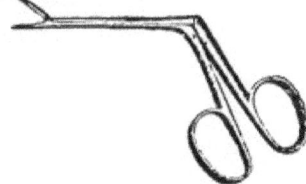

456   466   459

37

Eye Cases.

Case of Eye Instruments No. 1, contains: 1 plain wire
Speculum, 1 straight Iris Forceps, 1 Strabismus
Hook, 1 Scissors curved on flat, 1 silver Probe, 1
Spud for foreign bodies, 1 Daviel's Curette, 1 Beer's
Knife, 1 curved Needle, 1 Iris Needle, 1 Tyrell's sharp
Hook, fine Needles and Silk. In a neat morocco,
velvet lined case...................... ....$11 00

Case of Eye Instruments No. 2, contains: 1 Graefe's
Speculum, 1 Fixation Forceps, 1 Cilia Forceps, 1 Anel's Probe, 1 Agnew's Lachrymal Knife, 1 Desmarre's
Lid Scarifier, 1 Scalpel, 1 Strabismus Hook, 1 Strabismus Forceps, 1 Beer's Cataract Knife, 1 Graefe's
Cataract Knife, 1 curved Needle, 1 Cystotome, 1 Tyrell's blunt Hook, 1 Daviel's Scoop, 1 straight Kera-
tome, 2 angular Keratomes, 1 straight Iris Needle, 1
straight Iris Forceps, 1 straight Iris Scissors, 1 curved
Iris Scissors, 1 Bowman's stop needle, Needles and
Silk. In a fine morocco case, with knives in metal
racks $24 00

Case of Eye Instruments No. 3, W. F. Mittendorf's
Contains: 1 Mittendorf Eye Speculum, 2 Noyes Cata-
ract Knives, 1 curved Iridectomy Knife, 1 Agnew's
Lachrymal Knife, 1 Cystotome and Scoop, 1 Shell
Scoop, 1 Bowman's Needle with Stop, 1 straight
Needle and Spud, 2 Strabismus Hooks, 1 Lid Retrac-

tor, 1 Prout's Entropium Forceps, 1 Scalpel, 1 curved
Iris Scissors, 1 curved blunt pointed Strabismus
Scissors, 1 Fixation Forceps spring catch, 1 Cilia
Forceps, 1 curved Iris Forceps, 1 Prout's Needle
Holder, 1 set Bowman's probes, Needles and Silk, 1
Artificial Leech Cylinder with Piston. In a morocco
covered velvet lined case, Knives in a movable
metal rack.................................$32 50

Case of Eye Instruments No. 4, Agnews, contains: 1
pair Graefe's Specula, 1 Fixation Forceps spring
catch, 1 Desmarres double Lid Retractor, 1 Agnew's
Lachrymal Syringe with two pliable silver points, 1
set Bowman's Probes, 1 Bowman's Lachrymal Direct-
or, 1 Weber's Probe, 1 Agnew's Lachrymal Knife,
1 Prout's Needle Holder, 1 Scissors straight small
round points, 1 straight Scissors, 1 Scissors curved
on flat, 1 Agnew's Strabismus Hook with eye, 1 plain
Strabismus Hook, 1 Strabismus Forceps, 1 Tatooing
Needle, 1 Beer's Cataract Knife, 1 Graefe's Cataract
Knife, 1 Liebreich's Cataract Knife narrow, 1 Cystotome, 1 small Cataract Needle, 1 h. r. Scoop, 2 Stop
Needles, 1 Iris Forceps straight, 1 Tyrell's Sharp
Hook, 1 Iridectomy Knife angular, 1 Desmarres Para-
centesis Needle, 1 large Scissors for enucleation,
Needles and Silk. In fine hard wood polished in and
outside case.......................................$40 00

Case of Eye Instruments No. 5, Knapp's, contains: 2
Desmarres Lid Holders, 2 Fixation Forceps (spring
catch), 1 Artery and Needle Forceps, 1 hard rubber
Lid Plate, 1 Porte-caustic, 2 Bistouries, 2 Lance-
shaped Knives, 2 Knapp's Entropium Forceps (right
and left), Strabismus Scissors, 2 Strabismus Hooks,
1 set Bowman's Probes, 1 Weber's Cataract Knife,
1 Blunt Hook, 1 Cystotome, 1 silver (Knapp's) Cataract Scoop, 2 Knapp's Cataract Knives, 1 Beer's Ca-

taract Knife, 1 Stop Needle, 2 Sickle-shaped Needles,
2 Iris Forceps, 2 Iris Scissors, 1 Foreign Body
Needle, 1 Knapp's Foreign Body Hook, 1 Tatooing
Needle. In a fine morocco case, with Anel's Syringe
separate 47.00

Ear Instruments.

450	Applicator..	15
451	Curette Bucks blunt	50
	'' Ely's sharp..........................	60
452	Diagnostic tube	35
	Double Canula for polipy......................	50
	Ear Spout	25
453	Eustachean Catheter Hard rubber	25
	'' '' plated	50
	'' '' silver	90
	Faucial Catheter Pomeroys 1	00
454	Eustachean Vaporizers......................... 2	50
	'' '' Hartmann	90
455	Forceps Wildes...	75
456	'' Noyes 2	75
457	'' Roosa's 1	25
458	'' Hollers 1	00
	'' Pomeroys 1	75
460	'' Politzers 1	50
461	Furuncle Knife...............................	75
466	Hearing Horn London......................... 3	00
467	'' Fan (audiphone) 5	00
	Iodine Inhaler Buttles	75
	'' '' Pomeroys.....................	30
	'' '' Roosas 1	00
	'' '' Hartmanns	90
473	Mastoid Chisel Hartmanns 1	00
474	'' Chisel	50
475	'' Gouge	50
476	'' Drill, Bucks 3	00

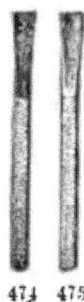

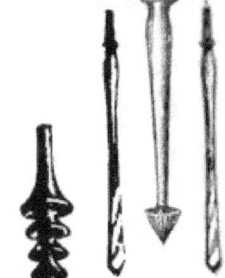

474   475   473   496   503   476

479

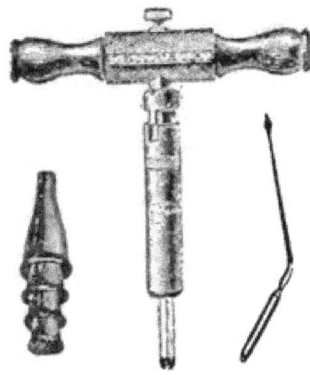
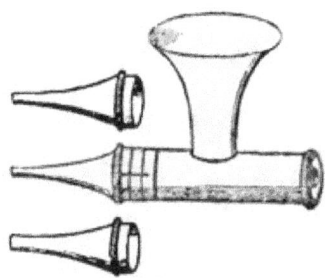

480   491

500   478   520   490

515

481

485   505   514   510   513

41

Mastoid Drill, Knapps   2 75

"   Drainage tubes   50

478   "   Drill, Wilsons...   3 50

479   "   Mallet, lead filled.   1 25

480 Otoscopes, Bruntons 4 50

481   "   Electric . ... .15 00

   "   Siegles 2 50

485 Politzer Bag complete   8 oz. red or black 1 50

"   "   "   10 os. .   ..   1 75

486 Probes Bucks, silver   35

" plated.................................   15

Pus Basins (see Amputatiug)

490 Scarifier Bacons........................ ....   4 00

491   "   Cup Bacons   40

496 Scoop double hard rubber   40

328 Scoop Gross........   50

500 Specula, Grubers set 4.   45

"   "   metal 1 00

503   "   Wildes Hard Rubber set 3......   40

Ear Cases.

Ear Case No. 1.—Contains: 1 plain Ear Mirror, 3 h. r.
Eustachean Catheters, 1 h. r. Ear Syringe, 1 set of (3)
Wilde's Specula, 1 Wilde's Ear Forceps, 2 Cotton

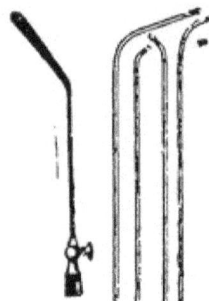

551 552 553 565

559

563 594

572

559 580 570 576 582 590 587

959

602

589 593 597 600

Fig.

Carriers, 1 Wilde's Polypus Snare, 1 Tympanum Perforater. In a morocco case, velvet lined $11 00

Ear Case No. 2.—Contains: 1 Ear Mirror (with headband and handle), 1 set of Politzer's or Wilde's h. r. Ear Specula, 1 set of Gruber's G. S. Ear Specula, 1 Politzer's Ear Forceps, 3 G. S. Eustachean Catheters, 3 h. r. Eustachean Catheters, 1 set Gruber's Tenotomes, 1 Wilde's Snare, 1 h. r. Ear Syringe, 1 Tympanum Perforator. In a neat morocco case$19 00

Ear Case No. 3.—Backs contains: 2 Curettes (blunt), 2 Curettes (sharp), 4 Cotton Holders, 1 silver Probe, 1 Mastoid Process Knife, 1 Port Acid Glass, 1 blunt pointed curved Bistoury, 2 Maryngotomes, 1 Furuncle Knife, 1 sharp pointed curved Bistoury, 2 Drills for Mastoid Process, 1 Mirror (3 inches diameter), 1 Wilde's Ear Forceps, 1 Blake's Snare (silver Canula), 1 set (4) Wilde's Specula. In a fine morocco case..$22 00

Nasal Instruments.

	Applicator, steel			15
	"	Bartons		20
551	"	Bosworths		60
552	"	Gleitsmanns Aluminum each		25
553	"	Goodwillies		75
	"	Robertsons (Chromic acid)		2 00
	"	Smiths		1 25
	"	Stucky		1 20
	"	Teets		25

Atomizers (see Throat.)

553 Canula Beloques plated.......1 00 silver....... 2 00

553 " double for removing Polypi 50

Chisel and Gouges Seilers..................... 4 00

" Hartmans, 3 sizes each.................. 1 00

" " for mallet 3 in set............ 4 00

563 Clamp permanent Bosworths.................. 1 25

45

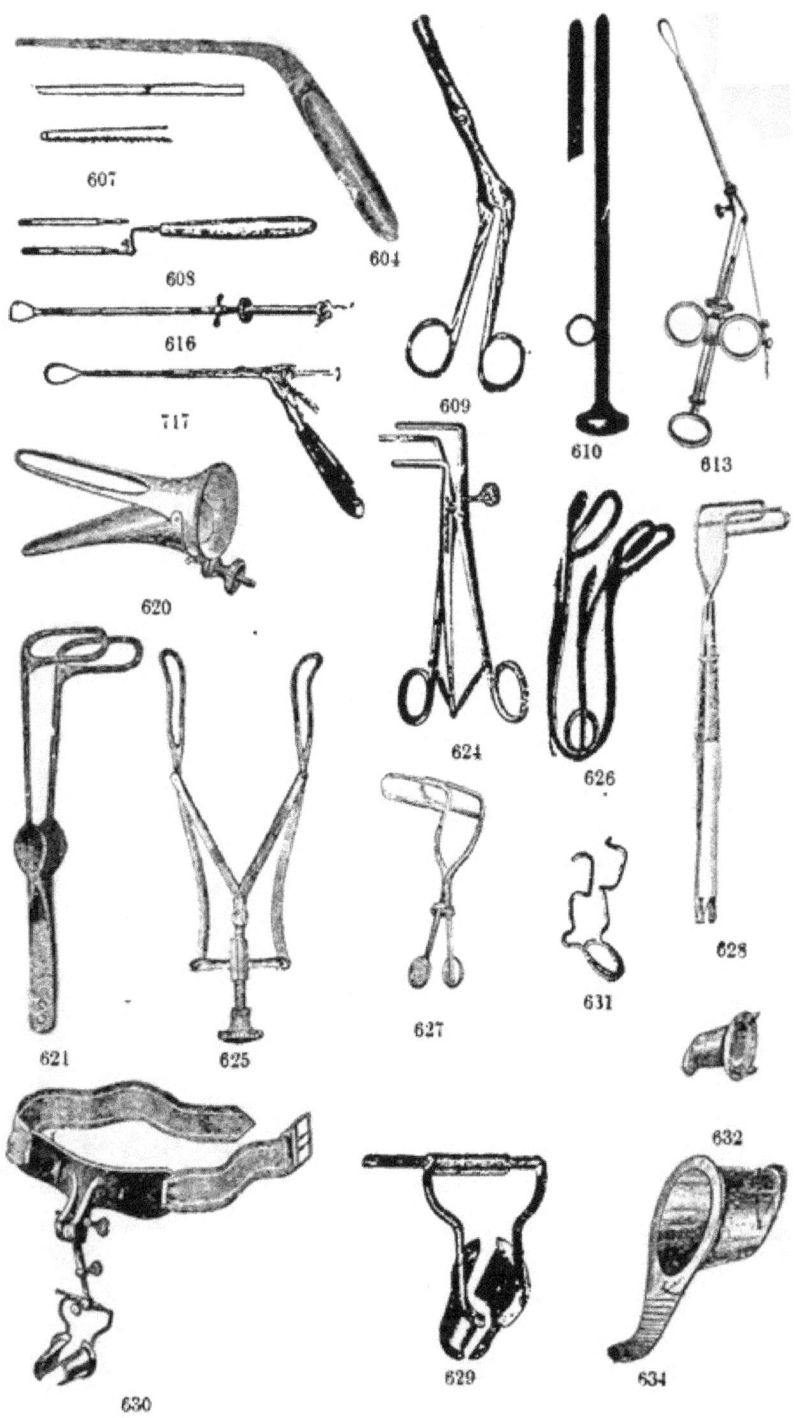

607

604

608

616

717

609

610

613

620

624

626

628

621

625

627

631

632

630

629

634

	Retractors, White's Uvula childs				$1 00
	"	"	"	adults	1 00
599	"	"	'	folding	1 50
600	"	"	"	Aluminum	2 00
	Rhinoscope Mirror				50
602	"	Duplays			2 50
	Rhinoplastic forceps Adams				2 25
	"	"	Bosworths		2 50
604	Saw, Bosworths, up or down cut				1 00
	"	Bucklins			2 25
	"	Curtis			1 00
607	"	Paine's			1 25
608	"	Rices			2 00
609	Scissors, Knight's				1 75
610	"	Smith's			3 00
580	"	Myles			4 00
	"	Seilers			1 75
	Septum Punch Roberts				4 00
	"	"	Sajous		10 00
	"	"	Steels		4 00
613	Snare, Bosworth, 2 tips				1 75
	"	Douglas			3 75
	"	Hopes			7 00
	"	Jarvis, single			1 25
	"	"	double		1 50
616	"	Sajou's			1 50
617	"	Wright's			6 00
620	Specula Bacons				00
	"	Collins			1 50
621	"	Bosworth			60
	"	"	plain		25
	"	Cohen's			1 50
624	"	Elsbergs			1 50
625	"	Frankels			1 25
626	"	Goodwillies			50

650

652

655

651

663

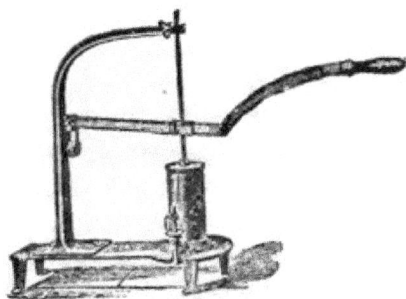

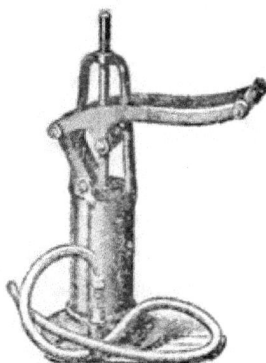

658

657

636

660

627 Specula Jarvis, Plain ....,.....$ 30
628   "   "   Operating 1 00
629   "   Myles.................................. 1 25
630   "   " improved 2 50
631   "   Palmers, 3 sizes @..................... 60
632   "   Simrocks............................. 60
      "   Thudichums.......................... 70
      "   "   with set screw.............. 1 00
634   "   Tubular 1 00
636 Syringe Post Nasal, hard Rubber 1 oz 75
      "   "   "   "   "   2 oz 1 25
      "   "   "   Warners 60
    Transfixion Needles Jarvis @.................. 30

Throat Instruments.

650 Air Receiver complete consisting of an 18 inch seamless
    copper nickel plated receiver tested to 120 lbs., air
    gauge, pump, silk covered tubing, cut off with bayo-
    net catches, and 3 glass spray tubes..........$32 00
    With 3 hard rubber, or 3 metal spray tubes...... 34 00

651 Air Receiver complete with nickle plated seamless Cop-
    per Receiver, Air gauge, pump, silk covered tubing,
    cut off with bayonet catches, 3 hard rubber Spray
    tubes on handsome finished Oak Table $36 00

652 Air Receiver with Lever pump fastened on a handsome
    finished Iron base with Air gauge, silk covered tub-
    ing, cut off with bayonet catches fastened on 3 glass
    Sass' Spray tubes, all parts of Pump and Receiver
    highly finished and nickle plated............$70 00
    Air pump, (see fig. 650) with T. handle all parts nickle
    plated $7 50

655 Air Pump extra long finely finished on Iron Base $10 00
        This is the best T. pump made.

657 Air Pump O'Learys Lever action $11 00

658 Air Pump, Lever action, finely finished, pumps to
    150 bls. never gets out of order. $21 00

660 Air Pump, Double Oscillating with gauge designed for exhausting as well as compressing air for atomizing purposes, and for compressing oxygen.

The construction of the pump is such that a pressure of 125 lbs. to the square inch can be obtained with little effort. The fly-wheel, frame part, and base are nicely japanned in black and ornamented in bronze, and all bright parts highly finished and nickle plated—making this pump especially adapted for the office.....................................$30 00

Without gauge 27 00

Air Tank only as Fig. 650 with gauge 18 00

663 Air Tank, made of steel, tinned in and outside, or Japanned. With gauge registering from one to one hundred lbs. and latest improved high-pressure valves warranted not to leak. Also, provided with couplings for attaching rubber tubing. Size, 10 inches by 32 inches...............................$20 00

L. & L. Automatic Hydraulic Air Compressor.

This device obviates the necessity of any of the larger pumps being worked by manual power. The difference in the price of each is but slight, and the Hydraulic Compressor accomplishes more automatically than the ordinary Hand Pump will with manual labor, it will give the same amount of pressure as the water system with which it is connected, and should be used in places where the water pressure is sufficiently high to produce the desired air pressure $25 00

670 Vaporizer Champion $4 00

672 Vaporizor Geyser, for vaporizing medicated Albolene, used with an Air Receiver....................$12 00

673 Vaporizer for purifying Air in the Sick Room can be used with Cresoline or any other Medicament.$2 50

Applicator, plated 25

  "   Silver. 75

672

673

670

677

676

675

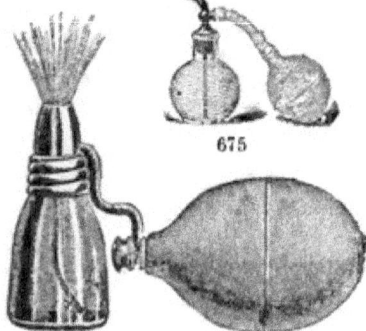

679

680

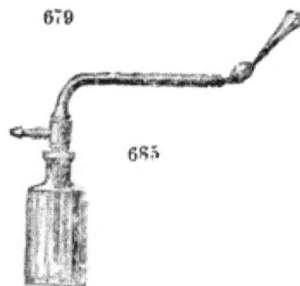

685

686

53

690

692

708

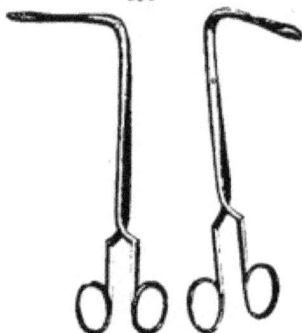

700

707

700

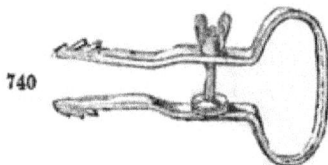

740

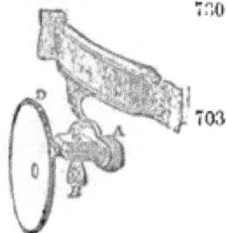

703

722

728

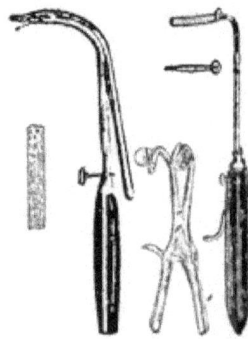

720

703 Head Mirrors, 2½ in. Nasal Rest or Pomeroy band . $2 00

  "   "   3   "   "   "   "   2 25

  "   "   3½   "   "   "   "   2 50

  "   "   4   "   "   "   "   2 75

707 Sardys improved head band only 1 00

708 Throat Mirrors, metal handles 5 sizes, each 50

  "   "   without handles 5 sizes, each 40

Universal Handles, ebony 25

  "   "   metal 50

Laryngoscopic Case, containing 3½ inch Head Mirror and Band, Tongue Depressor, 2 Cotton Applicators 2 Throat Mirrors, Universal Handle, set of 4 Hard Rubber Ear Specula in a velvet-lined Morocco case 6 00

720 Intubation Set, O'Dwyer's latest improved 20 00

722 Light Concentrator, McKenzie's 3 00

  "   "   "   improved 3 50

  "   "   "   "   with arm 6 00

725 Adjustable Gas Stand, extreme height of large size 48 inches 10 00

Adjustable Gas Stand, extreme height of small size 18 inches 6 00

For the greater convenience of those who use our improved Laryngoscope, we have constructed an Adjustable Gas Stand, in two sizes. The smaller size is for the desk or table, the larger to be placed on the floor, near the couch of the patient.

They are very strongly made, have a heavy iron base to increase steadiness and have a generally neat appearance. Iron supporting the burner is very readily raised or lowered after slightly loosening the centre screw; by tightening the same the burner support is held firmly in position and the joint is made gas-tight.

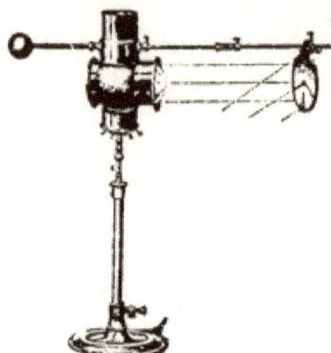

731   732   725

741

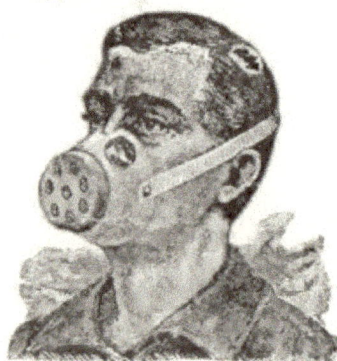

743

735   745

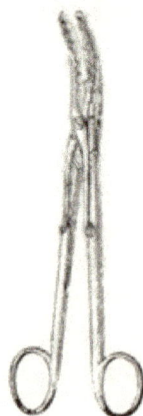

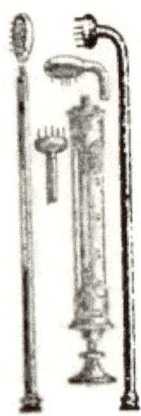

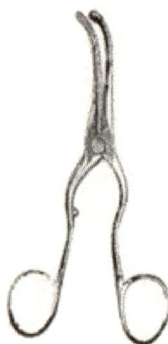

760   752   758   760   782

728 Adjustable Gas Stand with 6 feet, Gas tubing
    Goose neck and Shade $7 00
730 Adjustable Gas Bracket, polished brass 7 00
         "   "   "   nickel plated 7 50
731 Collin's Lamp 4 00
732 Adjustable Stand to fasten on table, with Students
    lamp, Mackenzies Condensor, Mirror and Ad-
    justable Mirror Arm 21 00
735 Inhaler Robinson's 50
    "   Blake's 2 50
    "   Tyndale's 3 00
    Mouth Gag, Denhardt's (see Fig. 720) 2 50
    "   "   Mussey's 5 00
740   "   "   Powers 4 50
    Powder Blowers curved, to slide, maroon bulb ... 75
741   "   "   "   with scoop, Pat 1 00
743   "   "   "   Robinson's universal 2 tips 1 75
    "   "   "   "   "   3 "   2 00
    "   "   "   Urban's pocket 2 00
    Probang Bristles 1 25
    "   Graefes 1 50
    Retractors, Trachea sharp, single 1 00
    "   "   blunt   " 1 00
    "   "   "   double 1 25
745   "   "   "   "   with elastic 1 00
    "   Uvula Leffert's 1 25
    "   "   White's 1 00
    "   "   "   folding 1 25
    Scissors throat 4 00
    "   Uvula 1 50
750   "   "   with claws 2 75
    "   "   Smith's 3 50
752 Seibert's tubes for injecting tonsils in cases of
    Diphtheria 2 tips, 3 50,   complete with   4 50
          4 " 5 00,   Syringe & case   6 00

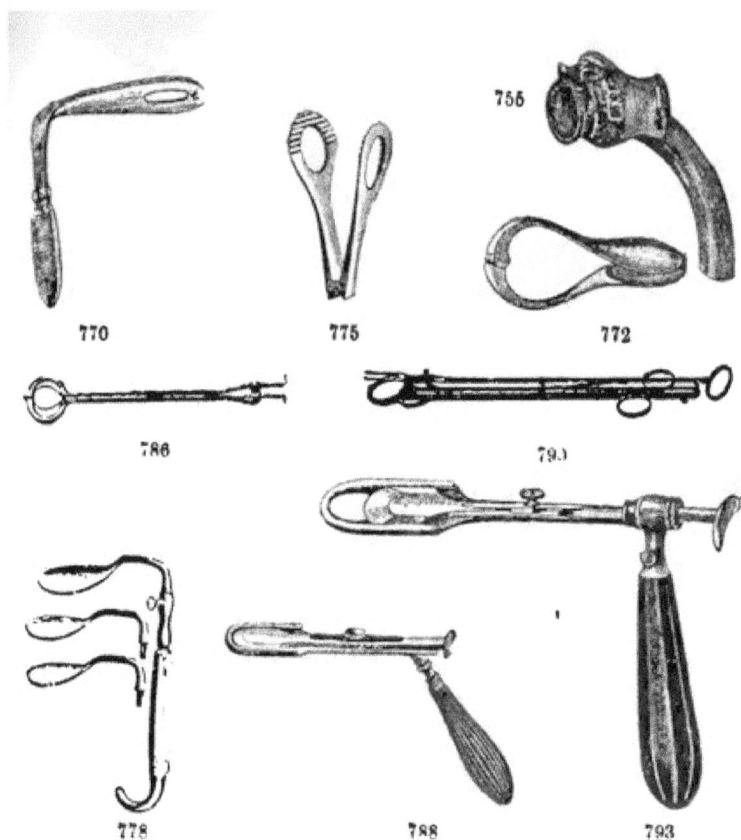

770   775   755   772

786   790

778   788   793

We are now making several new Electrical Instruments among others a **Head Light** to be used (with Storage or Cautery Battery) in place of the ordinary Head Mirror, for examining the Throat, Nose, Vagina, Urethra etc.

We also furnish complete **Electrical Outfits** for Throat and Nose Work.

Descriptive Circular on Application.

755	Trachea Tubes, hard rubber			$1 00
"	"	coin silver		4 00
"	"	aluminum		3 00
"	"	plated		2 50
"	"	forceps Gross		1 25
758	"	"	" Meuniers	2 50
"	"	"	" Bergmann's	3 75
"	Dilator Elsberg's			1 75
"	"	Leffert's		1 50
"	"	Luer's		1 50
"	"	v. Roth's		1 50
760	"	" Trousseaus'		1 75
	Tracheotome, Langenbeck			2 50
	"	Pithas		3 00
770	Tongue Depressor, Bosworth 0.50 Silverine			75
772	"	"	Goodwille's 0.50 Aluminum	1 00
"	"	Sass		65
"	"	Smith's		40
725	"	"	folding fenestrated	50
"	"	Tuerck's 1 blade 1 50 Aluminum		1 75
"	"	" 2 " 1 75 "		2 50
778	"	"	" 3 " 2 25 "	2 75
	Tonsil Bistoury, Tuerck's			1 25
782	"	" Conealed		2 50
"	Knife Tobold's			1 00
"	Electrode, Lincolns			2 50
786	"	" Snare Knight's		3 50
	Tonsilotome, Billing's			5 00
"	Fahnenstock's			3 50
788	" McKenzies			4 50
790	Tonsilotome Mathieus			$5 00
793	" Mandeville's revolving			6 00
"	Marconies			7 50
"	Sajous			7 50
"	Tiemann's			9 00
"	" 2 blades			13 50

59

Oesophageal Instruments.

Bougie, English, elastic cyl., $1.00; olive pointed$1 25
"   Bulbous, h. r. and whalebone stem 1 50
"   "   "   set of 6 .... 3 50
Bristle Probang, sponge tipped, $1.; ivory tipped 1 25
Graefe's Probang hinged Bucket at one end,
   Sponge at other end 1 50
Oesophageal Forceps, plain 1 75
"   "   Burge's 3 00
"   "   Fauvel's 2 00
"   "   Elberg's 5 00
"   "   Spiral 3 50
Stomach Tube elastic 1 00
"   "   " o'ive pointed............. 1 25
"   "   Rubber 30 inch............. .... 1 00
"   "   " 60 " funnel end 1 25
"   "   " 60 " with bulb........ 1 50
"   Pump Toswell's..................... 1 75
"   "   U. S. A. 10 00

Obstetric Instruments.

Obstetric Manikin for Demonstrating, leather
  covered, with Foetus and placenta 35 00
Schultze's Obstetric Illustrative Charts 16 00
Obstetric Forceps, Bedford 5 00
"   "   Budd 5 00
"   "   Burdicks 6 50
803 "   "   Elliots..................... 4 00
804 "   "   Hales..................... 3 50
805 "   "   Hodges 4 00
806 Obstetric Forceps, Lusk Simpsons.............. 4 00
"   "   McLanes.................... 5 00
809 "   "   McGillicuddys...............19 00
810 "   "   Simpsons 4 00
811 "   "   Sawyers............. .... 3 50
812 "   "   Taniers 12 00

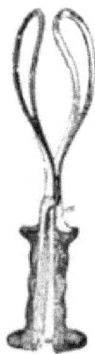

803

805

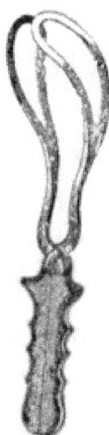

806

811

804

810

814

818

809

820

825

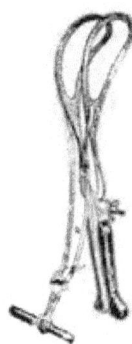

812

827

829

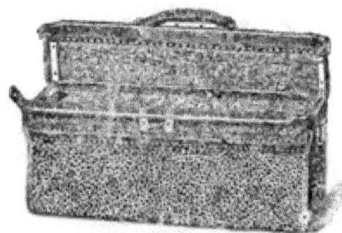

862

861

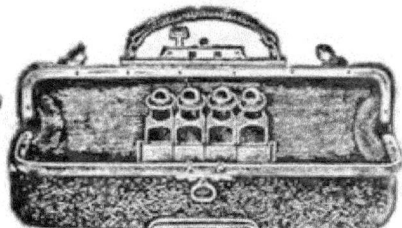

860

61

Pelvimeter Collins $6 00

    "   Kings............................ 3 00

    "   L. & L. 6 00

860 Obstetric **Instrument Bags.** Black Leather, lined inside with Chamois, containing pocket and four glass stoppered vials, 14 inch 3 00

                      15 " 3 25

                      16 " 3 50

                      17 " 3 75

                      18 " 4 00

We also have the above in black Alligator Leather add $1.50 to each price,

861 Obstetric Instrument Bags stiffened sides Black Leather lined inside, 13 inch 4 50

                      14 " 5 50

                      16 " 6 25

                      18 " 7 50

862 Instrument Satchel, Black Grain **Leather** 10 inch 4 00

                              12 " 5 00

                              14 " 6 00

                              16 " 7 50

                              18 " 8 00

Same in Alligator, add $1.50 to each price.

Hodge's Obstetric Pouch containing Hodge's Forceps, Blunt Hook & Crotchet, Perforator in a Canvas pouch........................ .... 8 00

Elliot's Obstetric Pouch containing Elliot's Forceps, Placenta Forceps, Blunt Hook and Crotchet, Craniotomy Forceps, Scissors, Perforator in a Canvas pouch...................12 50

Obstetric Pouch, Thomas' contains: 1 Elliot's Forceps, 1 Thomas' short Forceps, 1 Cranioclast, 1 Placenta Forceps, 1 Blunt Hook, 1 Crotchet, 1 Thomas' Trephine, 1 pair Scissors, 1 Naegele's Perforator, 1 Thomas' Cephalotribe, 4 Screw Cap

839

832   833   836   838   841

842

847

845

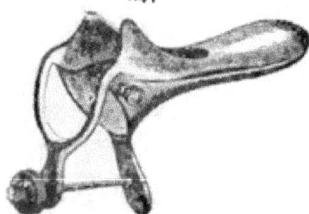

900

914

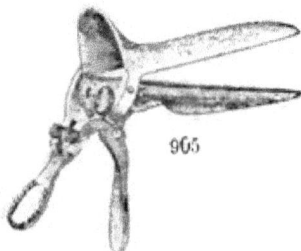

905

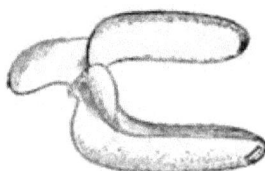

903

917

911

930

909

Bottles, in a Valise form case, with loops to hold
Instruments.......... ....65 00

Vaginal Instruments.

900	Vaginal Specula, Brewers		1	50
"	"	" Aluminum	2	50
"	"	Bathing hard rubber		75
"	"	" metal	1	00
"	"	Byrnes	7	00
903	"	" Cleveland's, with belt	2	25
"	"	Cusco s	1	50
"	"	" Aluminum	2	50
905	"	" " folding	1	75
"	"	" Aluminum	2	50
"	"	" Storer's	1	50
"	"	" Aluminum	2	50
"	"	Dawsons Erich's	9	00
"	"	" Sims	3	25
90.)	"	" Edebohl's	2	25
"	"	" Aluminum	4	00
"	"	Erich's Dividing blades	7	50
911	"	" Ferguson's glass		30
"	"	" hard rubber		75
"	"	" metal	1	00
914	"	" Graves	1	50
"	"	" Aluminum	2	50
"	"	" Asceptic	2	00
"	"	Goodell's	4	00
917	"	" Hales	2	00
"	"	" Aluminum	2	75
"	"	Hepburne's	1	25
"	"	" Aluminum	2	25
"	"	Higbees	1	50
"	"	" Aluminum	2	00
"	"	Howards	2	00
"	"	" Aluminum	2	75

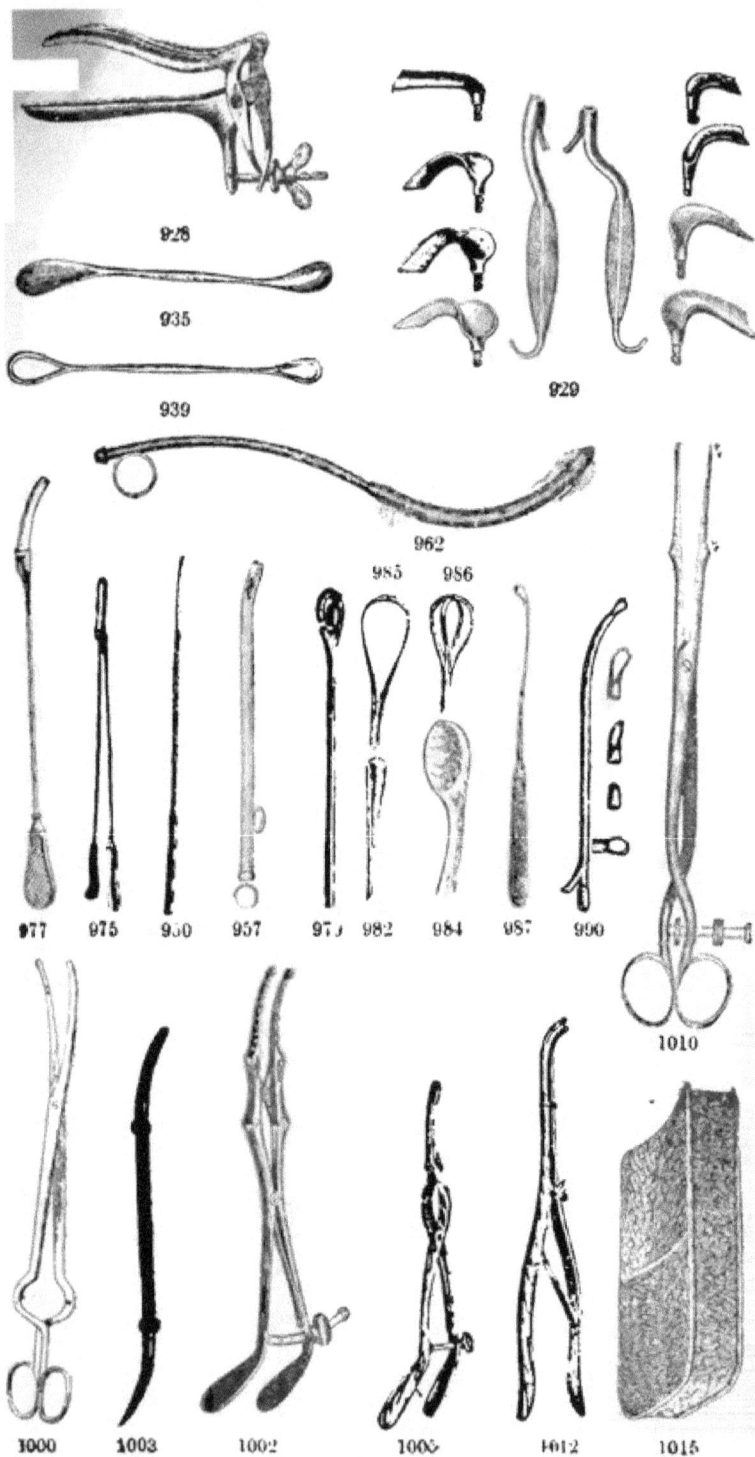

928

935

939

929

962

985   986

977   975   950   957   973   982   984   987   990

1010

1000   1003   1002   1006   1012   1015

66

Vaginal Specula, Hunter Erichs			$6 50
"	"	Jacksons	2 25
"	"	Leonards	2 50
"	"	Miller's	1 50
"	"	" Aluminum	2 50
"	"	Mundes	1 50
"	"	" Aluminum	2 50
"	"	Nelson's	1 75
"	"	" Aluminum	2 75
"	"	Nott's 3 sizes, each	1 50
"	"	" Aluminum	2 50
"	"	Simon's 5 in set 6 50 8 in set	12 00
"	"	Sims 5 sizes, each	90
"	"	" Aluminum	1 75
"	"	Storer's	1 50
"	"	" Aluminum	2 50
"	"	Thomas Cuscos	1 75
"	"	" Aluminum	2 50
"	"	" Sims	5 00
"	"	Taylors	1 50
"	"	" Aluminum	2 50
"	Depressors, Hunter		50
"	"	" Sims	50
"	"	Garrigues	1 00
"	"	Notts	50
"	"	Sims	40
"	"	Wylies	40
"	Dilators, Hanks set of 10		7 00
"	"	Sims glass 5 sizes each	25

Gynaecological Instruments.

Applicator, plated		25
"	Hard Rubber, Budds	25
"	Silver	75
"	Aluminum	50
"	Phillips	1 50

Applicator, Turner's with slide
    "   Sims with slide ....
957 Catheter, Female glass 0 15 plated 0 35 silver
      "   "   Sims glass 0.15 hard Rubber
      "   Chamberlins Glass..................
      "   Garrigues   "   ...., ...
      "   Ayer's, hard Rubber................
962   "   Bozemans double currant plated ....
      "   Fritchs   "   "   "
      "   Hollers   "   "   "
      "   Jennisons   "
      "   Kellys   "   "   "
      "   Luchesi   "   "   "
      "   Notts   "   "   " ....
      "   Skene or Goodmans self retaining ...
      "   " double Catheter ....
Caustic Holder, Hard Rubber ......
      "   "   with silver burner
      "   "   " Platinum burner
      "   Probes Lentes..... ..
975 Cervix Shield, Polks.... ....
      "   "   " don l .. ..
977 "   "   Wylie's .... //
Counter Pressure Hook, Emmet's ......
979 "   "   " Hank's.   '.... ...
Cupping Syringe Thomas ....
      "   "   Wylie's ....
Curettes, Uterine, Craigens .. ....
982 "   "   Hanks...................
      "   "   Mundes ....
984 "   "   Simons, 3 sizes each ......
985 "   "   Sims, 3 sizes each ....
986 "   "   Skeenes ....
987 "   "   Thomas 3 sizes each ......
      "   "   with douche ....
990 "   "   "   " Abbotts ......

1000	Dilators Uterine, Atlee's				$1 50
	"	"	Barnes set of 3		1 25
1002	"	"	Goodall's		6 50
1003	"	"	Hanks, per set of 10 sizes		2 50
	"	"	Leonard's		3 00
1005	"	"	" curved		3 50
	"	"	Notts		1 50
1010	"	"	Palmer's		2 00
	"	"	Sims		2 00
1012	"	"	Wylie's		1 75
	"	"	Waltons		4 00
1015	Douche pan, Hank's zinc				1 50
	"	"	" Agate		4 00
	Drainage tubes glass				15

Drainage Tubes (see Page 15).

	Forceps, Dressing, Bozemans plain			1 50
1020	"	"	Elliot's without catch	1 25
1021	"	"	" with "	1 50
	"	"	Harvey's	1 50
	"	"	Mundes	1 50
1024	"	"	straight	1 25
1025	"	"	curved with catch	1 50
1027	"	"	Sims	1 50
	"	"	Thomas	1 50
	"	Haemostatic, Lawson Taits		75

Forceps Haemostatic (see Page 7.

	Forceps Hystorectomy Clamp Delerys			2 50
	"	"	" Polks	2 25
	"	"	" Thomas	2 25
	"	Pedicle Spencer Wells straight, obtuse or angular each		2 75
1033	"	" Nelatons		3 00
	"	Tissue Sims straight, plain		1 25
	"	" " curved		1 50
	"	" Thomas slide catch curved		2 75

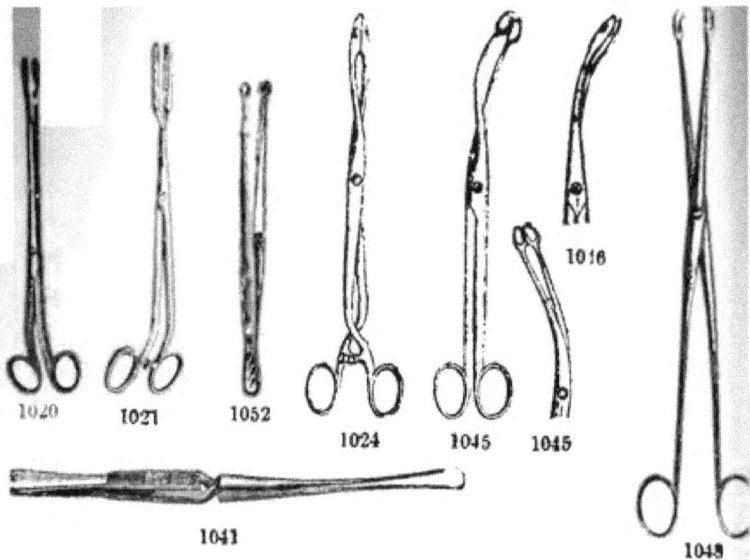

1016

1020   1021   1052   1024   1045   1045   1048

1041

1027

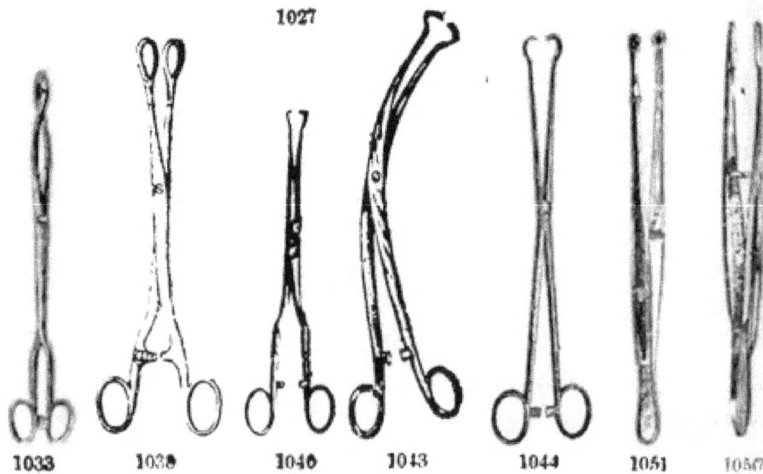

1033   1038   1046   1043   1044   1051   1057

1054

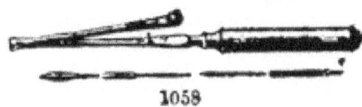

1058

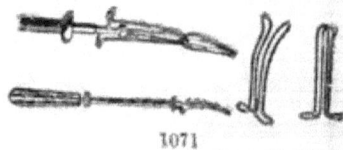

1071

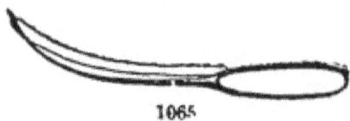

1065

	Forceps Tissue Thomas' straight....		$2 50
1038	" Sponge, Hunter's		2 00
	" " Wylie's		2 00
1040	" Tenaculum Edebohls...............		2 50
1041	" " Notts....................		2 00
	" " Hank's		1 50
1043	" " Skene's		1 50
1044	" " Wylie's		1 50
1045	" Vulsellum 2 pronged		2 00
1046	" " angular		2 25
	" " 4 pronged angular		2 50
1048	" " Hank's....................		1 50
1050	" Wire Twister Sim's....................		1 50
1051	" " " Thomas'		1 50
1052	" " " Emmet's................		1 75
	Hysteotome, Atlee's		4 50
	" Bishop's.........................		6 00
	Intra Uterine stems (see Pessaries).		
1054	Knife Boldts.................................		1 50
	" Edebolds double edge		1 25
	" Sims revolving		2 25
1058	" " " with 4 blades...............		5 00
	" Scalpel sharp..........................		1 00
	" " blunt......................		1 00
	Knot tier Ayers		4 00
	" " Carrol's		1 50
	Needles Cervix, per dozen............ ...		50
	" " Emmets triangular.............		60
	" " Hanks..........		75
	" " Polks.....		75
	" " Schnetters .		75
	" " Sims round		50
	" " Hagadorn's		75
	" Pedicle, Wylie's 3 curves each		1 00
	" Perineum Ashton's		75

Needles Perineum Dickinsons $1 25
" " Gazzams 1 25
" " Jacksons 1 25
. " Hollers 1 25
" " Peaslee's straight 75
" " " curved 75
" " " set of 3 .... 1 25
" " " Skene's 1 00
1065 " " Goodwillie's 1 50
Needle Holder, Crosby's 2 00

Needle Holder's see page 10.

Overiotomy Clamp, Atlees 4 00
" " Esmarchs............... 4 00
" " Lawson Taits.. . ... 9 00
" " Spencer Wells 4 50
" " French 4 50
" Trocar, Emmet's 3 50
" " Tait's................. . 12 00
" " Wylie's 2 50
1071 Outerbridge's Introducer, asceptic handle 2 00
" Wire Dilator each...... 25
Pouches Wylie's made of Canvas 1 50
Probes Sim's 30
" Aluminum 50
" Silver............................ . 75
" Plated 25
Repositor, Campbells 18
" Carrol's.................... 2 25
1080 . Elliots 2 75
" Emmets... 3 00.
" Gardners...... .... 4 00
" Guernseys.... 75
1084 " Sims........ ...... 3 50
" Skenes.... ....·. 4 00
" Wylies 3 75

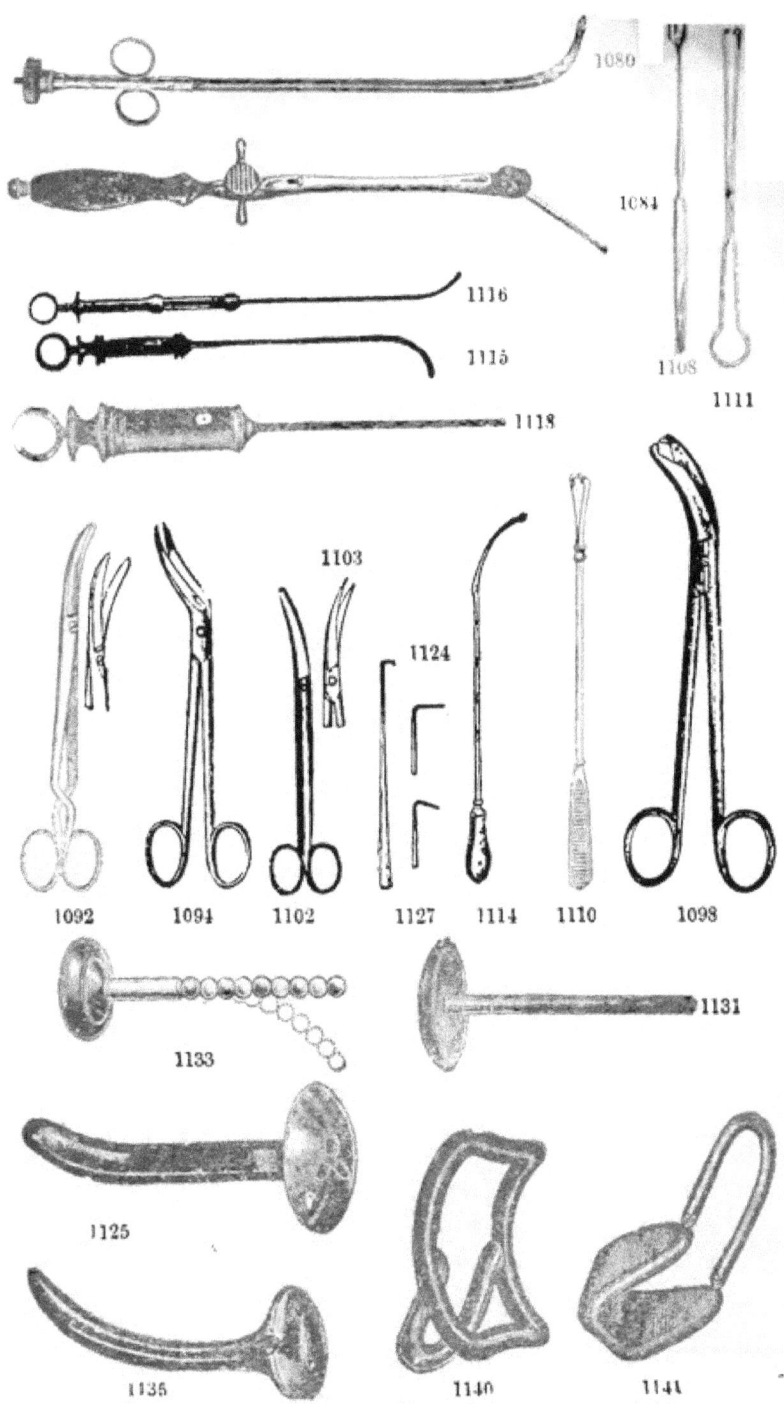

1080

1084

1116

1115

1108

1111

1118

1103

1124

1092   1094   1102   1127   1114   1110   1098

1131

1133

1125

1135   1140   1141

Revolving Knife Sim's see **Fig. 1058.**

1033	Sac Forceps Nelatons		$3 00
"	" Thomas		2 25
	Scarifier Buttles		60
"	" w. Tenaculum		1 40
	Scissors, Bozemans		3 00
1092	" Boldts		1 50
"	Clarks		5 25
"	Dawsons		3 00
1094	" Emmets right or left		2 75
"	" full curved		2 75
"	" angular		1 75
"	Urethral Button Hole		5 00
1098	" Hanks, Diamond		2 75
"	Heywood-Smith's		7 00
"	Jenk's Perineal		3 00
"	Kuechenmeisters		2 50
"	Pratts		2 00
"	Sims', Angular		2 00
1102	" " Curved, Blunt Points		1 50
1103	" " " Sharp Points		1 50
"	" Straight, Blunt Points		1 25
"	" " Sharp		1 25
"	Skene's Button Hole		4 00
"	" Hawk Bill		6 00
"	Wylies wire cutting		2 50
"	Scoop Thomas serated,		2 00
1108	Shield Sims		50
	Skewers Sims		50
	Speculum Barnes		2 50
"	Urethral Folsoms		1 00
"	" Skenes		1 25
	Sponge Holders straight		35
1110	" " curved		35
1111	" " Hussons		50

1143   1144   1148   1149   11.4

1160   1195   1197

1176   11.5   11.0   1190   1193

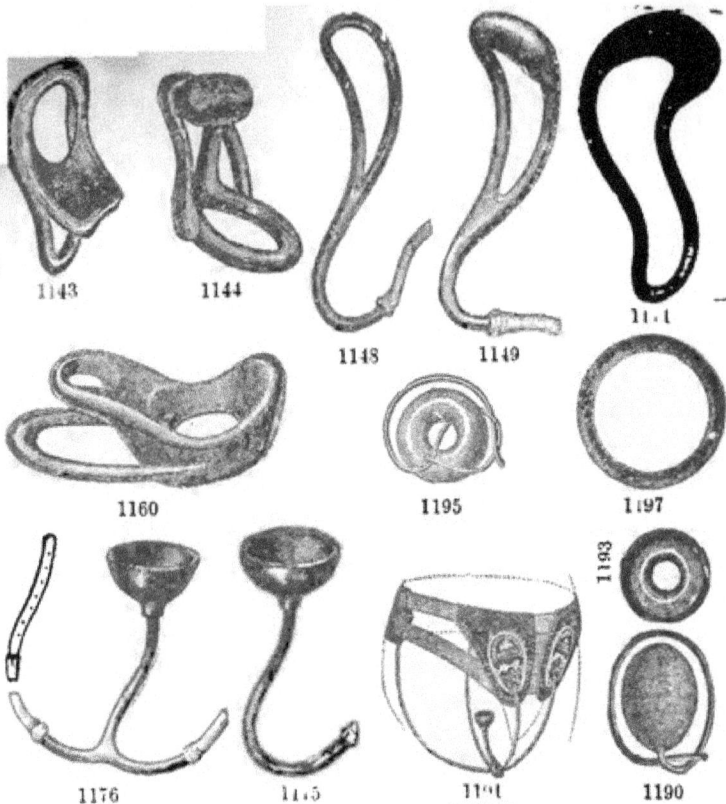

The Perfection Operating Table.

1198

We also handle the following Tables and Chairs,
Circular and Prices on application.
Dagget Table, ...
Childs Chair, ...
Harvard Chair, ... :. . .
Perfection Chair, ...
Yale Chair, ...
We are now making a Combination set of **Stirrups**, for
examining Patients on the Table, Bedside, or for Operating
use similar to the Edebohls Stirrups Price complete $8 00

Pessaries for Anteversion.

	Anteversion, h. r., Thomas' 1st pattern			$	60
1140	"	"	" 2d "		60
1141	"	"	" 3d "		60
	"	"	" 4th "		75
1143	"	"	" 5th "		60
1144	"	"	" 6th "	. 1	50
	"	"	" 7th "	. 1	50
	"	"	" 8th "	. 1	75
	"	"	" 9th "	1	50
1148	"	"	Cutters with band	1	00
1149	"	"	" Thomas with band	1	25
	"	"	Gehrungs		25
	"	"	Hurds hard rubber	1	00
	"	"	McIntosh with supporter	2	50
	"	"	Pallens	1	25
	"	"	Wylies		50

Pessaries for Retroversion.

	Retroversion h. r.	Bow		20
	"	"	Chamberlains		75
	"	"	Coles	2	00
	"	"	Cutter w. belt	1	00
	"	"	" Thomas	1	25
	"	"	Fowlers	1	25
1160	"	"	" w. bow	. 1	75
	"	"	Gehrungs		40
	"	"	Hanks	1	00
	"	"	Hewitts		45
	"	"	Hodges		20
	"	"	Horseshoe		20
	"	"	Mc Intosh w. belt	2	50
	"	"	Noegeraths		40
	"	"	Pallens	1	50
	"	"	Smith h. r.		20

Retroversion Smith soft rubber .... $ 60

    "   " Thomas Smith soft rubber... 75

1171   "   "   "   " h. r......... 40

  "   "   " Hewitts 60

Pessaries for Prolapsus.

Cutters, h. r. with ring and belt 1 00

1175   "   "   " cup "   " 1 25

1176   "   Thomas with cup and belt 1 50

  "   Donaldsons pliable with belt 1 50

Cushiers with belt......................... 2 50

Concave, h. r. .... $0 25............glass 20

Globe,   " .... 0 60............ " 30

1190   "   soft rubber inflatable 50

1191 Mc Intosh improved with Supporter 2 50

O'Learys with screw or spring and belt.. ..... 2 25

1193 Soft rubber ring Inflated................... 30

  "   "   "   " heavy............... 30

1195   "   "   " Inflatable 40

  "   "   " Pear shape 50

1197   "   ring, whalebone rubber covered.... 25

  "   " copper rubber covered 25

  " hard rubber 25

  " soft metal................ 25

Zwangs, h. r..... .... 1 00

1198   Table Examining Perfection.

No. 4—Best Antique or Sixteenth Century Finish,
quartered oak, full cabinet, four drawers,
three zinc trays, nickel-plated trimmings,
hand carvings, leather top cushions and pillow 80 00

No. 6—Antique Oak or Sixteenth Century Finish,
quartered oak, leather top cushions, nickel-
plated trimmings, full cabinet, without zinc
trays or hand carvings 75 00

No. 8—Same as No. 6, with embossed tannette
cushion and pillow... ....70 00

1204 1206

1205   1202   1208   1210   1214   1231

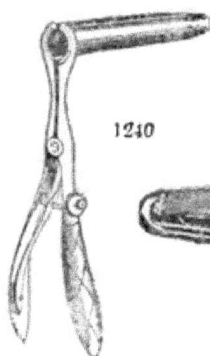
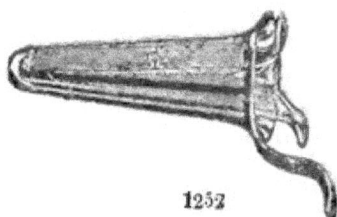
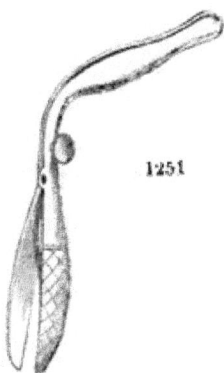

1240

1251

1252

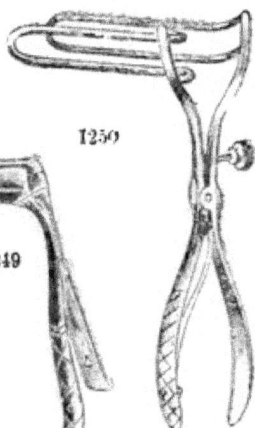

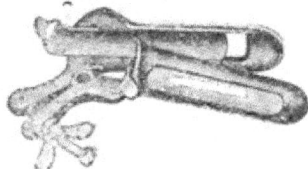

1250

1247

1249

212

Send for Circular of new Paquelius Cautery.

No. 10—Same as No. 6, with artificial leather
cushions and pillow.................... $65 00

No. 12.—Best Antique or Sixteenth Century
Finish, quartered oak, leather top cushions,
without cabinet or drawers, but with nickel-
plated trimmings and hand carvings 60 00

No. 14—Same as No. 12, with leatherette cushions and pillow, without hand carvings.... 55 00

Table Dagget's Examining.

Standard, wood top 60 00

  " upholstered in imitation leather...... 65 00

  "   " in leather 70 00

Star, wood top 30 00

" upholstered in imitation leather... .... 35 00

Cushions, canvas....................... 5 00

  " imitation leather................. 10 00

  " leather........................ 20 00

Table Columbia Examining.

Table made in oak with drawers and stirrups .. 30 00

Table without drawers including stirrups. 20 00

Head Pillow in Leather 2 00

Rectum Instruments.

1200	Bougies elastic, cylindric			75
1201	"	" conic.......		90
	"	" olive		90
1202	"	Bulbous on whalebone stem.		1 25
	"	" 6 sizes on whalebone stem..... ..		2 75
	"	Wales, 1, 2, 3 each..		1 25
		4, 5, 6		1 35
		7, 8		1 50
		9, 10		1 65
		11, 12....		1 75
1205	"	set of 6 sizes, hard rubber		2 00

Urethral Instruments.

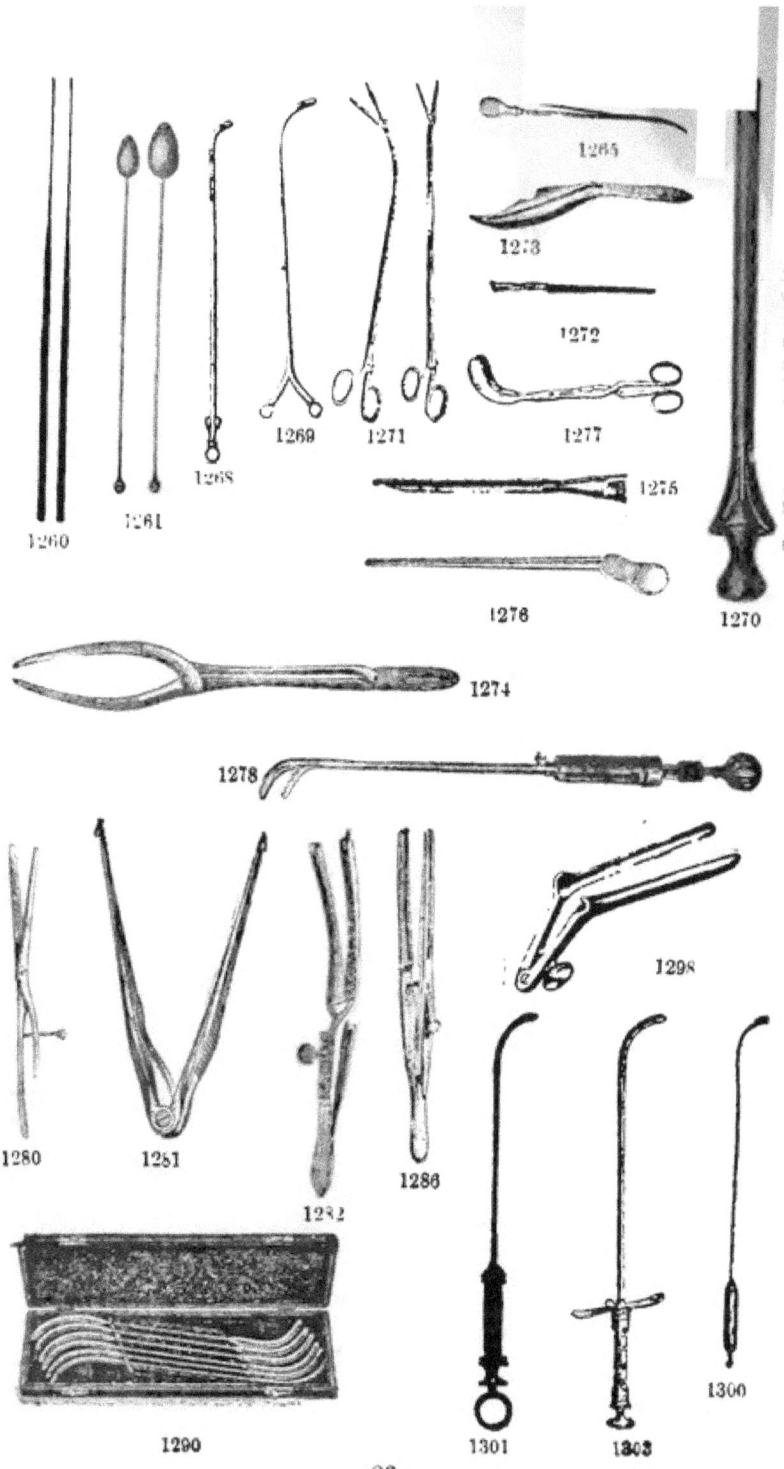

1265

1273

1272

1260 1261 1268 1269 1271 1277 1275 1276 1270

1274

1278

1280 1281 1282 1286 1298

1290 1301 1303 1300

Hernia Syringe Warren's 6 75
" Syringe Tube, Warren's 1 50
Herniotomes, Allis'. 5 25
" Warren's 4 50
1274 Lithoclast 4 00
" Gouleys............................ 4 00
Lithotomy Bisector Woods 3 00
1275 Lithotomy Bistourie, Blizards.......... 1 25
" " Gross's 1 25
" " Littles................. 1 25
" Crutch, Clover's 6 00
" " Hupps.................. 8 00
" Director 1 00
1276 " " Little's 1 00
" Forceps, Gross 3 50
" " Gouley's................. 1 75
1277 " " Little's 2 00
" " Markoe's................. 2 00
" " Curved 1 75
" Gorgeret 2 00
" Scoop, Luers 1 75
" Staffs, Lateral 1 25
" " Littles..................... 1 25
" " Markoe's.................... 1 25
1278 Lithorite Bigelows latest..................... 25 00
" Key's 26 00
" Thompson's 23 00
Lithotrite Evacuator, Bigelow's 9 00
" " " latest 20 00
" " Otis 22 50
" " Thompsons 11 00
1280 Meatotome Civial's 3 75
" Gouley's 2 00
" Knife Otis 1 00
" " Piffards 1 00

Ointment injecting canula silver, King's 2 50
" " screw cap, hard rubber, Kings 1 50
This instrument is so arranged that medicated
Vaseline can be forced in the Urethera. The
instrument is put up complete in case with
2 salve boxes for ointment 5 00
1281 Phymosis Forceps, Girdner's 1 50
1282 " " Henry's.................... 2 50
" " Knox...................... 3 00
" " Lewis' 2 00
" " Ricords................... 2 00
1286 " " Skillen's 2 00
Powder Blower, Waechters.................... 1 50
" Applier, Sages 10 00
1290 Sounds Steel, Fowlers double curved 6 in case .. 5 00
" " Van Buren's nickel plated each... 50
" " Wyeth's straight 75
" " " curved.................. 1 00
" " set of 12, in Morocco Case 8 00
" " " " Oak Polished Case.... 9 00
" Piffards fossal 50
" Van Buren's cupped.................. 90
" Cooling Winteritz plated................ 1 50
Spermatic Ring 60
Speculum Glass............................. 45
" Folsoms, or Otis................... 75
1298 " Caro's 1 00
" Skene's 1 50
" Tuttle's 1 75
Stone Searcher, Andrew's 1 75
1300 " " Thompson's 2 50
1301 Syringe Bumstead's in case................... 1 50
" Ultzmanns 1 25
1303 " Keys deep Urethral............... 3 00
" tube Keys plated 1 00

     NOTE.—Prof. J. A. Wyeth's Urethrotome, Straight
     and Curved Sounds, Bulbous Bougies, Wha-
     lebone Filiform Bougies, Skewers, &c., are
     made according to his latest designs.

Miscellaneous Instruments.

1350 Aspirator small, with Trocar, Aspirating Needle,
     2 hypodermic needles, and Stop cock in
     Morocco Case 2 25
1351 Aspirator L. & L.
     This Aspirator possesses advantages over any other in
the market, and its low price places it within the reach of
every practitioner. The pump has two automatic metal
valves, which enables the operator to quickly change it
from an exhaust to a force pump, thus converting it into an
injector. The valves are easily cleaned and not injured by
coming in contact with liquids, and as the fluid passes only
through the needle, long pipe, end stop-cock, the instrument is easily cleaned. It is nickle plated and put up in a
neat velvet lined case with 3 Needles, 1 Trocar and 16 oz
Bottle,
          Price 10 00

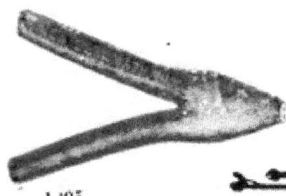

1306

1305

1316

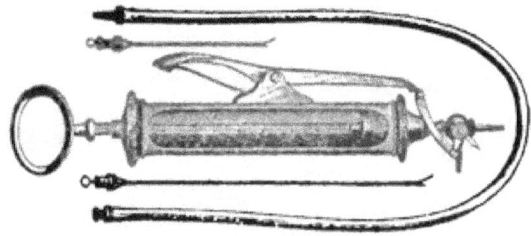

1314

1318

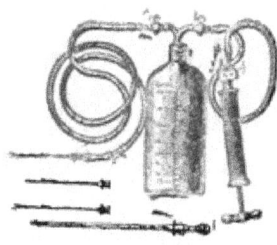

1350

1351

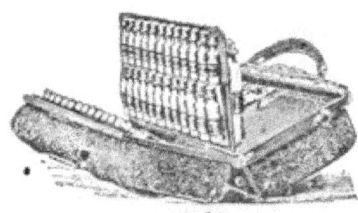

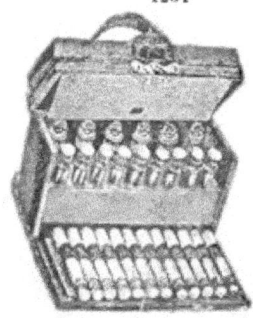

1476

1366

89

Aspirator same as above without bottle (stop cock has rubber cork to fit any bottle) in neat Case, price 7 50

1353 Aspirator Peaslee's, with 2 needles, 1 Trocar and canula 8 00

Aspirator Phelps, with 2 needles, 1 Trocar and 1 Phelps Trocar........................ .15 00

Hypodermic Syringe, glass barrel, metal fittings with finger rest, 2 needles in case 1 50

Hypodermic Syringe as above with 1 or 4 bottles, in metal Case............................ 2 00

Hypodermic Syringe as above with 1 or 4 bottles, in Aluminum Case 2 50

Hypodermic Syringe as above in morocco case with 6 Tablet Vials 2 00

1360 Hypodermic Syringe, glass barrel with our patent piston, which can *never dry out*, with 2 needles in morocco case 1 75

Same in hard rubber case........... 2 25

    " " metal case 2 25

    " " " " with 4 pellet vials. 2 25

    " " Aluminum case............. 3 00

    " " morocco case with 6 pellet vials 2 25

1360

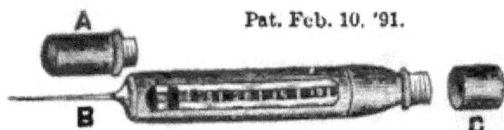

Pat. Feb. 10, '91.

A

B

C

1364

1364 Hypodermic Syringe, Pencil Case pattern being the only Syringe that can be rendered perfectly Aseptic.

Fig. A small box to hold Tablets. B needle which
is carried in the piston rod at C.

By putting a few drops of carbolized oil in the piston
rod it will always keep the packing oiled as well as the
needle Aseptic,

Price 2 00

With 2 tablet boxes A 2 25

" 3 " " A .... 2 5?

All our Syringes are made so that a pellet can be drop-
ped in the syringe.

1366 Buggy Case, size, 9¾x6¾x2 inches.

Contains 24-2½ drachm, 7-6 drachm. and 5-3
drachm wide mouthed corked vials. Also a
box for hypodermic syringe. Best black seal
leather 8 50

Doctors often complain that the leather loops holding
the bottles in medicine cases are a source of trouble and
inconvenience. The bottles are often too loose or too tight
because these loops are either too large or small, and
besides this, the face of the bottle is so covered that little
room is left for the label.

But case No. 1366 is one to which none of these ob-
jections can be made. A Physician's *multum in parvo* case,
very compact and substantial, and of attractive appearance.

The vials are held in by nickel tongues projecting over
the base of the vials, which are thus secure but can be
taken out without the slightest trouble. Physicians using
tablets, pills or pavules will find this case particularly
handy, as the bottles have mouths of good width.

1373 Buggy Case, size 10x7x4¾ inches.

Containing 6-1½ oz., glass stoppered, 9-1 oz., and
12-4 drachm corked vials. In the top is a box
with lid, size, 9x3¾x1, with instrument loops,
space large enough for a small sized stetho-
scope. The case closes with a spring catch,
and has a lock and key. It is most substant-
ially made, and will bear much rough usage10 50

NOTE.—In place of the 12-4 drachm vials, we can furnish
24-2 drachm vials, or 6-4 drachm and 12-2 drachm.
Please specify the number desired.

1375 Buggy Case, Size, 10 inches long, 6½ inches high, 4¾ inches thick.

Black morocco, contains 20-1 oz., and 13-6 drachm corked vials. Its corners and edges are protected with nickel trimmings, and as its leather is the best it will, with proper care, last for many years. It contains two covered boxes, each 4½x5½ inches and 1¼ inches deep, in which may be carried pocket instrument case or sundry other articles. Spring lock with key....11 00

NOTE.—This case can also be furnished to contain 7-6 drachm and 12-2 drachm vials, in place of the 13-6 drachm.

1376 Buggy Case, size, 10x7x6 inches.

Satchel Case, 10-inch frame; contains 18 1 oz., 11-6 drachm and 11-4 drachm corked vials, ample loops for instruments, box for sundries and a neat plush-lined case for hypodermic syringe 11 75

We think this is the most convenient satchel case in the market. All of the pieces holding the bottles are removable, transforming it into a traveling or instrument satchel.

SPECIAL LIST OF BUGGY, HAND AND POCKET Vial Cases sent on application.

Dissecting Cases.

Dissecting Case No. 1, Contains: 1 Scalpel, 1 Cartilage Knife, 1 pair Forceps, 1 Tenaculum, 1 pair Scissors, 1 Blow Pipe, Hooks and Chain in mahogany case 1 45

Dissecting Case No. 2, Contains: 2 Scalpels, 1 Cartilage Knife, 1 pair Forceps, 1 Tenaculum, 1 pair Scissors, 1 Blow Pipe, Hooks and Chain in mahogany case 1 85

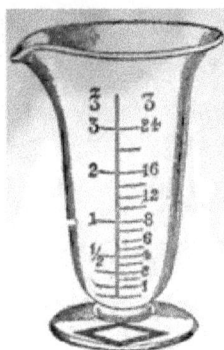

1383

1384

1385

1400

1878

1396

1389

1390

1392

1393

1384 Instrument Cabinet in Antique Oak............15 00
      Height less railing 40 inches.

Instrument Cabinets, Allisons and the Perfection
   Circular on Application.

1385 Instruments Cabinet for Dentist.

We desire to call attention to our new style of Dentist
Cabinet. It comprises four small and three large drawers,
a revolving compartment with four movable shelves which
can be turned outward (see second shelf with cups), thus
bringing the instruments conveniently within reach of operator,

Lower part of Cabinet is large enough to hold Basin,
Towels, etc. By closing and locking revolving compartment,
the four opposite drawers are securely fa tened. Altogether
a very desirable piece of Furniture for office use·

We make them in Antique Oak and Walnut. Mountings
are nickel plated Boxwood casters.

   Height of Cabinet to railing, 63 inches; width,
     28 inches ; depth, 16 inches 25 00

1389 Microscopes and Acessories, prices on Application.
   Operating Table Edebohls, with plate glass top .90 00

1390 Operating Table Foersters 35 00
      "   " Hirschbergs with plate glass top 100 00

1392 Prescription Scales Troemers $3\frac{1}{2}$ inch pans Ca-
     pacity 4 oz 8 00

1393 Presciption Scales Army
       $5\frac{3}{4}$ inch Beam, 2 drams to $\frac{1}{2}$ grain..... 3 50
       7   "   "   2   "   " $\frac{1}{2}$ " 5 00
       8   "   "   2   "   " $\frac{1}{2}$ " 6 00

Spit Cups glass white or blue 25
Tape Measure.........,3,   5,   6 ft.
   plain linen 0.30 0.35 0.40
   best linen 0.60 0.85 1.00
   Steel 1.25 1.50 1.65

1396 Tooth Forceps 40 differnet patterns plain 1.00 best 1 50
   We furnish all kinds of Dental Instrument prices
    on Application.

Rubber Goods.

Bandages see page 13.

Bed pan round. 2 75
"   "   "   with outlet Tube 3 50
"   "   "   "   "   oblong 4 00

Cushion Invalid
    9, 10, 11, 12, 13, 14, 15, 16, 18 in.
    1.25 1 50 1.65 1.85 2.00 2.15 2.25 2.50 2.75

Head Bag 6¼x9½ in. 30
"   "   6½x11 in. 40

1400 "   "   screw cap 6,   7,   9,   11 inch.
           0.65 0.80 0.90 1.25

Ice Bag. 7,   8,   9,   10,   11,   12, inch.
       0.50 0.60 0.65 0.75 0.85 1.00

Nursery Sheeting, $\frac{3}{4}$ wide per yd 70
         $\frac{4}{4}$ "   "   " 90
         $\frac{5}{4}$ "   "   " 1 20
         $\frac{6}{4}$ "   "   " 1 50

Rubber Tubing $\frac{1}{16}$ $\frac{1}{8}$ $\frac{3}{16}$ $\frac{1}{4}$ $\frac{5}{16}$ $\frac{3}{8}$ 1 2 $\frac{5}{8}$ $\frac{3}{4}$ 1 in.
    white .2½ .4 .6 .7½ .9 .11 .14 .18 .22 .25
    black or red .3 .5 .6 .9 .12 .15 .20

Syringes Vaginal No. 1 75
"   "   No. 2 . . . . 1 00
"   Davidson No. 1 . . . . 1 15
"   "   "   No. 2   . . . . 1 25
"   "   Molesworth 1 25
"   "   Lady's 2 25
"   Fountain   1,   2,   3,   4 qt.
          1.15 1.25 1.35 1.50

Urinals male and Female send for *Circular.*

Water Coils Abdominal
    7,   9,   11,   13,   15 inch.
    1.50 2.00 2.25 2.65 3.00

Water Coils for Head small 2 25
"   "   "   "   large 2 75
"   Coil Spinal 3 00

Diagnostic Instruments.

Aesthesiometer, Birdsalls ..		2	50
" Carrolls		2	50
" Hammond		2	00
" Peckhams		3	25
" Vance		3	50
Cyrtometer. Flints		6	00
Dynamometer		6	00
Percussor, Flints			60
" " h. r.			50
" Traubes		1	25
" Winterichs		1	00
Pleximeter, Flints			20
" Glass			40
" Ivory			75
Sphymograph, Dudgeons		18	75
" Ponds		30	00
" Mareys		40	00
Spirometer, Barnes		8	00
1425 " Shepards		25	00
Stethoscopes, Camman's		1	50
" " with spring		1	50
1428 " " folding		2	50
1429 " " University		1	75
" Dennisons		4	00
" Knights		3	00
" Pauls Vacuum		2	50
" Snowdens		1	75
" Snoftons		1	75
" Wilsons		2	50
" Cedar wcod			50
1436 " Hawksleys			75
1438 Thermometer, with indestructable register			75
" " " " colored bulb	1	00
" " " and certificate	1	25

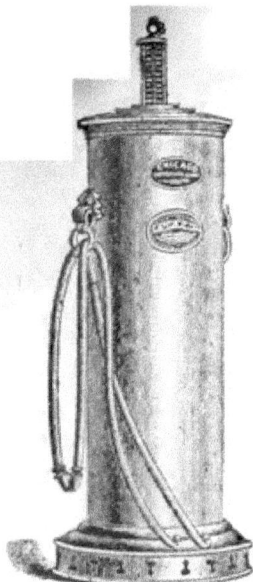

14.5

1428

1429

1436

1441

1438

1444

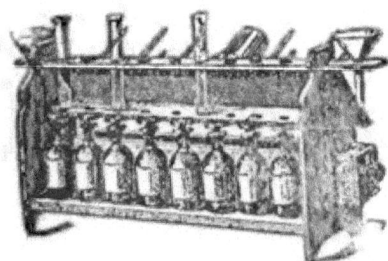

1448

1450

1447

1449

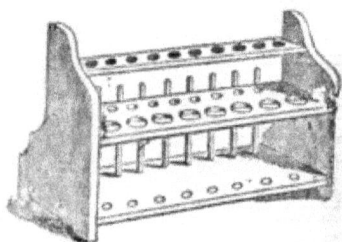

1481

1441 Thermometer, indestructable magnifying 1 25

  "   "   "   Hick's 2 00

  "   "   "   " minute .. 3 00

1444 "   Cases gold plated with chain add. 25

Urinary Examining Instruments.

Sachrometer Fermentation, Einhorns.......... 1 50

Spirit Lamp .... 1, 3, 8 oz.

         0.20 0.30 0.40

1447 Urinometer, plain with receptable 50

  " best "   " 1 50

  " hard rubber " 2 75

Uria Tube, Doremus.......................... 60

  " "   " and Thermometer combined 1 25

1448 Urine Test Stand, Luchesi No. 1, consisting of
ten Test Tubes, two Glass Funnels, Alcohol
Lamp, eight glass bottles for reagents, Urino-
meter, red and blue Litmus Paper, Filtering
Paper, two Watch Crystals, Stirring Rod and
Drop Tube 3 50

1449 Urine Test Stand, Luchesi No. 2:

 The case is made of hard wood, finely polished, and is
constructed on a new principle which possesses many advantages. The upper part, which forms the test tube rack
when in use, can be closed down and fastened; the hollow
slats holding the funnels slide into the case; the drawer
can be returned to its socket, thus forming a neat, compact
box that guards its contents from breakage and protects
them from dust, light and air. It contains 8 Reagent Bottles, 2 Glass Funnels, Alcohol Lamp, 2 Beakers, Porcelain
Evaporating Dish, 2 small Glass Evaporating Dishes, 14 Test
Tubes, assorted sizes, Test Tube Holder, Urinometer, Grad-
uated Pipette and Litmus Paper. Two of the reagent bottles contain the two solutions now used in making Fehling's
Test; one a solution of Sodio Potassic Tartrate, the other a
solution of Cupric Sulphate, both so adjusted that, with the
directions that accompany the case, a physician may easily
make both a qualitative and quantitative analyses for Gluc-
ose, thus enabling the practitioner not only to detect sugar

in suspected cases but to make comparative tests, from time, in known cases of Diabetes. They are always ready for use and will keep for any length of time.

Price of Case, with bottles all filled 7 00

1450 Urinary Test Tubes per doz. 40

1451 " " " Stand 50

" " " " with 18 Test Tubes... 1 00

Abdominal Supporters.

Abdominal Supporter, made of Satin Jean 3 00

Abdominal Supporter, made of Thread Elastic,
$5 00 to $7.00 of Silk Elastic $7 00 to 10 00

1500 Abdominal Supporter, Teufels System I or II
Satin Jean............................. 5 25

1501 Abdominal Supporter, Teufels System I or II
Satin Jean, White Horsehair 7 50

Abdominal Supporter, Teufels for Laparotomy,
System III Satin Jean 6 60

Abdominal Supporter, Teufels for Laparotomy,
System III Satin Jean White Horsehair...... 8 85

Abdominal Supporter, L. & L. for Laparatomy,
with hard Rubber pad 3 50

1505 Abdominal Supporter, Umbilical Teufels 6 50

" " " L. & L........ 3 50

" " " " w. Spring 3 00

" " " childs pure gum 80

" " " " w. Spring 1 25

" " Uterine with Cup Pessary 2 50

Elastic Stocking.

1510 Elastic Stocking for leg and thigh, silk......... 8 50

1511 " " " " " " thread 5 75

" " " " " knee " 3 50

" " " " " " silk 5 00

" " " " to garter " 2 75

" " " " " thread 1 75

" Anklet, thread....................... 1 50

" " silk 2 00

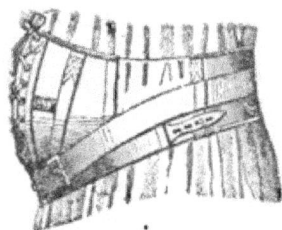

1500

1501

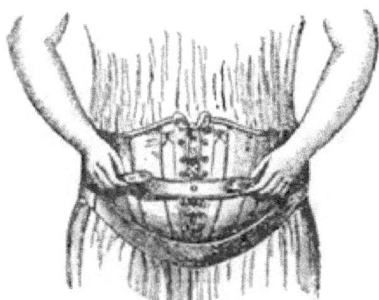

1505

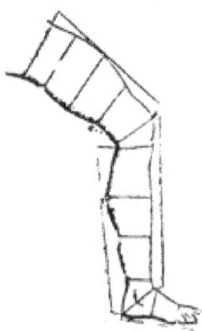

1510

1511

Fig. 2.   Fig. 1.

1512

A   B

1525

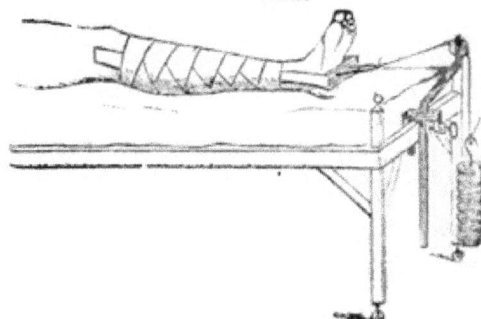

1530

Elastic Knee Cap, silk 1 75
"   "   "   thread 1 50
"   Calf piece   " 1 50
"   "   "   silk...................... 2 00
1512 "   Wristlet   " 1 50
"   "   thread 1 00
Measurements required see Fig. 1510, 1511 or 1512.

Shoulder Braces.
Gents' elastic................................ 1 50
"   steel back............................ 2 00
"   "   "Sure Cure" 3 00
Ladies' elastic, with skirt supporters 1 25
"   steel back,   "   " 2 50

Suspensory Bandages.
Suspensory Plain Cotton ......$0 25.......best   50
"   Silk................. 0.75.......best   85
"   " heavy 1 00
"   Cotton, with perineal straps......... 50
"   Silk   "   "   "   85
"   " heavy   "   " 1 00
"   Cotton with tape in bag............. 1 00
"   Silk,   "   "   " 1 25
"   Millianos for Varicocele 1 25

Trusses.
1525 Truss German, single calf covering 2 00
"   "   double...................... 3 75
"   French, fine single 2 00
"   "   " double 4 00
"   " ball and socket, single, fine calf skin 2 50
"   "   "   "   "   double   "   "   4 00
"   " hard rubber single 2 50
"   "   "   "   double.............. 4 50
"   Self adjusting h. r. single 2 50
"   "   "   "   double.............. 4 50
"   German calf covering Childs single 1 25
"   "   "   "   "   double....... 2 25

Fig,

Truss Umbilical see Abdominal Supporters.

To Measure a patient for a truss pass a tape
around the Patient between the Cest of Illium
and Trocantor Major, state Right, Left, or
Double, Scrotal, or Inquinal Hernia.

Crutches.

Crutches Common, per pair	2 25
" with paded Arm pieces per pair	2 50
" " " " " & rubber bottoms	4 00
" " " " " " "	5 00
" " spring arm pieces and patent rubber bottoms	6 00
" " do. do. Rosewood	9 00
Arm Sling Patent Leather	3 00
" " Wire gauze	2 75
Extension Apparatus Bucks	2 50
1530 " " Levis	5 00
Fracture Box	5 00
Head or Bed Rest	4 00
" " " with Arm Rest Attachment	5 00

Humane Restraints for the Insane.

Straight Jackets	10 00
Lynch Buckle	1 25
Same with waist strap	1 75
Canvas Muff	3 25
" Mit	3 25
Bed Strap	6 00
Mitts, with waist strap and patent lock buckle per pair	5 00
Anklets, with buckle per pair	4 00
Wristlets, " " " "	3 75

Invalid Bedstead Crosby's.

Bedstead	25 00
Head Rest	5 00
Excelsior Mattress	2 50
" " soft top	4 50

Splints.

Splints of Basswood per doz. 75

    "   "   "   lined per doz. 1 00

    "   Day or Pratts Wood, per set 37 50

    "   "   "   "   Double Inclined Plane

                                  $3.00, 3.75, 4 50

    "   "   "   "   Extension Bars 3.75, 4 50

    "   "   "   "   Ankles, r. and l., each

                          0.55, 0.75, 0,95, 1 15

    "   "   "   "   Patella, back plain

                        0.60, 0.70, 0.75,   90

    "   "   "   "   "   jointed or screw

                        1.50, 1.70, 1.90, 2 00

    "   "   "   "   Joint Arm with Screw

                        1.50, 1.65, 1 90

    "   "   "   "   Condyle and Humerus

                        0,55, 0.75,   95

    "   "   "   "   "   "   " jointed

                        0.75, 0,95, 1 15

    "   "   "   "   Forearm

               0.30, 0.40 0.45, 0.55, 1 15

    "   "   "   "   Forearm Squire's, jointed

                           0 75, 1 50

    "   "   "   "   Radius, curved, r. and l.,

                  each 0 25, 0.45, 0.55,   65

    "   "   "   "   Dressing, per set of five...   40

    "   Wire for entire Leg..................... 4 50

    "   "   " Forearm 2 00

    "   "   " Leg below Knee............... 2 75

    "   "   " Foot and Ankle 2 00

Splints Ahls or Johnstons, made of Felt.

Splints included in a Regular Set:

1) Inferior Maxillary ....Adult $0.60 Child's.   30

2) Clavicle ....   "   0.90   "   60

3) Shoulder Cap ....   "   0.90   "   60

4) Humeral..... ....Adult $0.50 Child's 30
5) Elbow, Right and Left each   "   0.90   "   60
6) Radial,   "   "   "   "   "   0.90   "   60
7) Ulnar   "   "   "   "   "   0.90   "   60
8) Femoral ....   "   0.90   "   50
9) Knee, Anterior, Right & Left "   1.20   "   90
10) Knee, Posterior,   "   "   "   1.20   "   90
11) Tibial, Right and Left....   "   1.20   "   90
12) Fibula,   "   "   " ....   "   1.20   "   90
13) Straight Pieces for Fingers and Toes.....each   30
14) Club Foot, Right or Left ....   "   90

Complete Sets, 50 pieces 25 Adult's and 25 Childs
   Price per Set 26 20

Laced Splints for Knee 6 00
   "   Corsets for Spinal Curvature 25 00

Splints Levis Metalic.

1) Radial, right and left, 2 Adult, 2 Child each.. 1 00
2) Elbow Joint, right and left, 2 Adult 2 Child
   each 1 50
3) Humeros, right and left, 2 Adult, 2 Child each   50
4) Phalanges, 3 in set each   15
5) Clavicle, 2 in set each   75
6) Maxilla, 2 in set each.................... ..   75
7) Femur, 2 Adult, 2 Child, each   50
8) Patella, right and left, 2 Adult. 2 Child, each. 1 50
9) Tibia & Fibula, right and left, 2 Adult, 2 Child
   each 1 00
The above set consists of 21 pieces in a neat Case
   Price 15 00
Stretcher, N. S. Hospital pattern 8 00
   "   U. S. A.   "   " 9 00

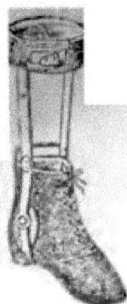

1532

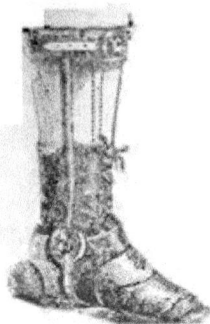

1533

1550

1535

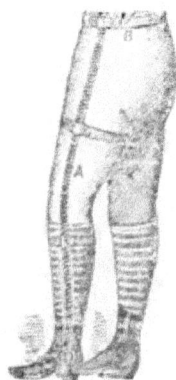

1540

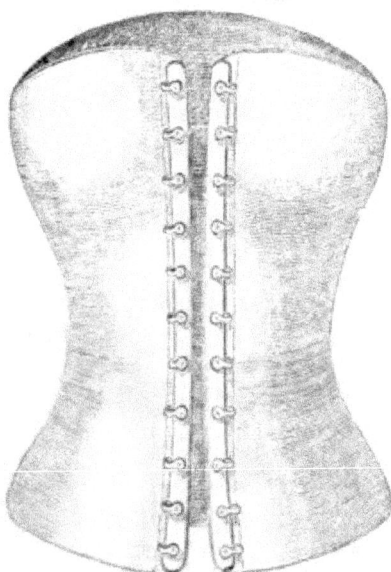

1554

1536

1538

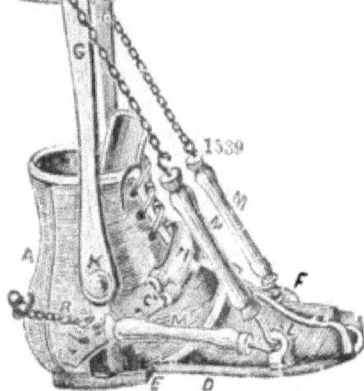

1539

1540

Orthopaedical Apparatus.

Ankle Extension Splints SEE SPLINTS.

Apparatus for Weak ankle single, spring, each

2.75 to 4 50

1532 Apparatus for Weak ankle double spring each

3.00 to 6 00

1533 Apparatus for Weak ankle double with leather

Tug 4.00 to 7 00

MEASUREMENTS REQUIRED.

1) Send a well fitting shoe to lace.
2) Length from Sole to Ankle Joint.
3) Length from Sole to Garter.
4) Circumference of Garter.
5) Right or Left.
6) Inclination of Ankle inward or outward.

1535 Apparatus for Bow Legs Single bar each $4.00 to 6 00
1536   "   "   "   " Double bars each 5.00 to 8 00
    "   "   "   " Double bars to lengthen between Ankle and Knee joint add to each 1 00

MEASUREMENTS REQUIRED.

1) Send a shoe to lace.
2) Circumference of Thigh midway between Perineum and Knee Joint.
3) Circumference of Leg below Knee.
4) Length of Sole to Ankle.
5)   "   "   "   " Knee Joint.
6)   "   "   "   " convexity of Leg.
7)   "   "   "   " middle of Thigh.

Apparatus for Club foot, Gross's each..$9.00 to 12 00
1538   "   "   "   " Detmolds each 4.00 to 6 00
1539   "   "   "   " Sayres each.. 7.00 to 10 00

MEASUREMENTS REQUIRED.

Place the sole of the patients foot on paper and sketch it, and give
1) Length of sole of foot,—inches.
2) Circumference of Garter,—inches.

3) Circumference above ankle.
4) Length from sole to Garter.
5) Right or Left foot?
6) Talipes—varus, or vulgus.

Dr. G. Doyle's Spiral Rotators. For Aftertreatment of Clubfoot and Correction of Inverted or Everted Feet.

Upright of Brace consists of a Spiral Spring. This rotated more or less before shoe is put on exerts a power upon leg or foot in opposition to rotation of the spring. Price, single...........12 00
" both legs........18 00

MEASUREMENTS REQUIRED.

1) Circumference of Body between crest of Ilium and Trochanter Major.
2) Circumference of middle of Thigh.
3) Circumference of Garter.
4) Length from Sole to Ankle Joint,
5)   "   "   "   " Garter.
6)   "   "   "   " Knee.
7)   "   "   "   " point to circumference (1).
8) Right, Left or Double.

Eclipting Spring for Flat Feet per pair......... 2 25
Send a Strong lace shoe and plaster cast of foot.
  Elevation.—Made of steel, japanned, the leather
sole riveted to the under side of the steel sole ... 4 50
  Leather stiched to steel sole..........$5 C0 to 7 00

Made of steel, nickelplated, the leather sole stitched to the under side of the steel sole $6.' 0 to 8 00

Elbow Apparatus Strohmayer's for Real and False Anchilosis of Elbow Joint. Extension in this apparatus is made by a screw. If desired it can also be made by means of ratchet, as in Sayre's Splints. Price, with Screw........$.0.00 to 14 00
" with Key and Ratchet 14.00 to 18 00

MEASUREMENTS REQUIRED.

1) Lenght from top of Shoulder to Elbow outside.
2) "   " Axilla to Elbow inside.
3) "   " Elbow outside to Wrist.
4) "   " Ell ow inside to Wrist.
5) Circumference of Arm above Elbow.
6) "   " " below Elbow.
7) "   " " at Wrist.

1550 Jury Mast, Sayre's......................... 7 00

MEASUREMENTS REQUIRED.

1) Lay a thin and flexible strip of lead along the
spine, moulding it exactly to the spine and all
its sinuosities from the op of head to the
middle of sacrum. With pattern trace carefully
the shape of the spine and head on a sheet of
paper, marking the points opposite the upper
and lower borders of the scapulae and the
crest of ilium, also the affected parts of the
spine. Send this tracing to us. Upon receipt
of application, we will mail free of charge to
parties desirous of ordering one of the above
named apparatus, a flexible strip of lead to aid
in securing the tracing above referred to.
2 Circumference of head, under the chin and
over the head.
3) Circumference of head, around the chin and
back of the neck. In taking this measure care
should be exercised that the tape encircles
these parts in an even horizontal line.
4) Circumference of body at crest of ilium,

Jacket made of Aluminum, price.......50 00 to 75 00
1554 Jacket made of Felt price 20 00
  "   " " Leather, price..................30 00
  "   " " **Paper**, price..........25.00 to 40 00
  "   " " Wood 25 00
  "   " " Plaster of Paris...............15 00
1558 Dr. L. A. Sayre's Apparatus for suspending
Patients while applying the Plaster of Paris Jack-
etc, ... 5 00
    We are prepaired to furnish e.ther of the above
Jackets but would recommend either the Wood

Fig

or Paper Jacket, the latter supersedes anything
heretofore contrived in lightness, durability and
imperviousness to moisture. In order to make
it we must have a Plaster of Paris Jacket, made
as well as possible, as the Paper Jacket cannot be
made to fit any better than the former.

Directions for making Plaster Cast on application.

Knee Extension, see Splints.

1560 Knee Joint Apparatus Strohmayer's....16 00 to 20 00
1562 Knock Knee Brace ....20 00 to 30 00
Measurements required same as Doyle's Rotator
send lace shoes.

Partial or Entire Paralysis Apparatus for one Leg
   " " "   "   plain ....12 00 to 16 00
   " " "   " with stop joints 16 00 to 20 00
   " " "   " for both Legs. 22 00 to 35 00

MEASUREMENTS REQUIRED.

1. Send strong shoes to lace completely down to
the toes. with low heels.
2. Circumference of Body between crest of Ilium
and Trochanter Major.
3. Circumference of Thigh one-quarter below
Perineum.
4. Circumference of Thigh one-quarter above
Knee Joint.
5. Circumference of Leg at Knee Joint.
6. Circumference of Garter.
7. Circumference of Ankle.
8. Distance from Sole to Ankle.
9. Distance from Sole to Knee Joint.
10. Distance from Sole to Trochanter Major.
11. Distance from Sole to crest of Ilium.
12. Right, Left or both Legs. Entire or partly.

Splint for Ankle Extension Phelp's .....6 00 to 10 00
Splint for Ankle Extension Sayre's.....12 00 to 16 00

MEASUREMENTS REQUIRED.

1. Trace the outlines of the Sole of the Foot on
a piece of paper.

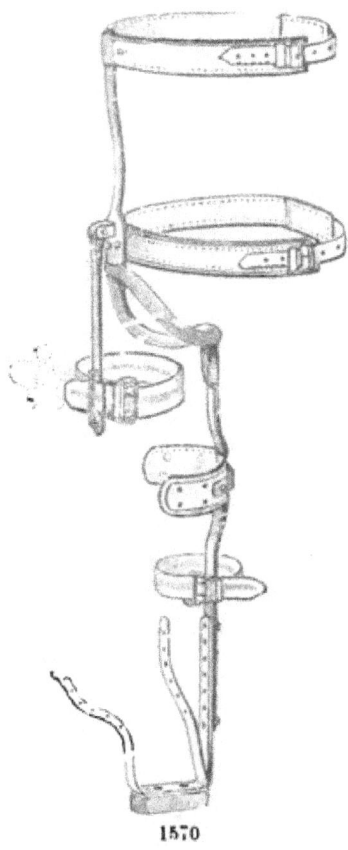

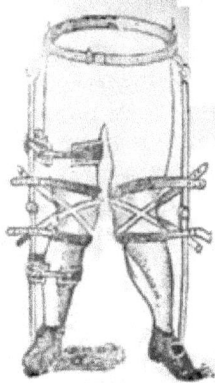

1572

1562

1575

1570

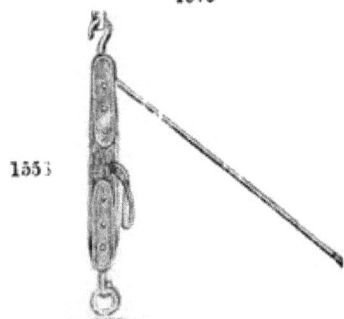

1553

1578

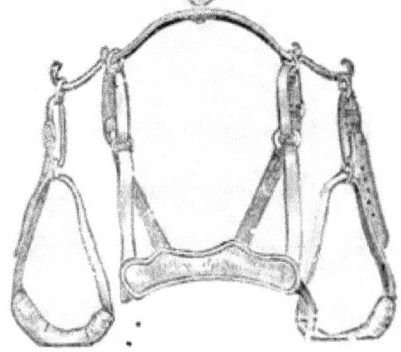

111

2. Right or Left Ankle.
3. Length from Sole of Foot to Garter.
4. Circumference at Garter.
5. Circumference of Ankle Joint.

Splint for Hip Extension, Judson's, plain...... 17 00
Splint for Hip Extension Judson's with key and
ratchet 24 00

1570 Splint for Hip Extension, Phelp's
This Splint is designed to absolutely immobilize
the joint and relieve intra-articular pressure.
Fixation and rest allow the processes of repair
to take place uninterupted by the trauma of
motion. Traction relieves intra-articular pressure and controls muscular spasm.

Hip Splint, plain, not polished 16 00
Same, polished and nickel-plated 22 00
Steel Elevation for well leg, according to finish
price, ... 3 00 to 6 00
Cork Elevation for well leg.................... 6 00
Wood   "   "   " 1 50

MEASUREMENTS REQUIRED.

1. Sex, age and weight of patient.
2. Right or left.
3. Circumference of Chest under Axilla.
4. Circumference of Waist.
5. Circumference between Iliac Crest and Troch-
    anter Major.
6. Circumference of Thigh at Perineum.
7. Circumference of Leg, middle of Thigh.
8. Circumference at Knee.
9. Circumference of Leg between Ankle and
    Knee Joints.
10. Distance between Axilla and Iliac Crests.
11. Distance between Iliac Crests and Trochanter
    Major.
12. Distance between Trochanter Major and
    middle of Thigh.
13. Distance between Perineum and middle of
    Knee Joint.

14. Distance between Perineum and Sole of foot ▓▓
15. Send lace shoe for the well foot.

Splint for Hip Extension, Sayre's, short 11 00 to 13 00

1572   "   "   "   "   "   long . .

With Plain Joint 16 00

With Platform Joint and Abducting Screw 23 00

With above and Rotation 27 00

Any of above with Spring Box, add 3 50

Any of above with Automatic Lock Knee Joint,

    add 6 00

MEASUREMENTS REQUIRED.

1. Circumference of Body between crest of Ilium
   and Trochanter Major.
2. Length from the above point to Sole.
3. Length from the above point to Knee.
4. Right or Left.

Splint for Hip Extension Taylor's 16 00

Splint for Knee, Hutchinsons 10 00

For making this we require a wrapping of the
    limb, of Plaster of Paris Bandages, from which
    we make a facsimile in Wood or Paper.

Splint for Knee Extension Phelps plain $7.00 to 10 00

Splint for Knee Extension Phelps Nickle plated

                              $10 00 to 13 00

1575 Splint for Knee Extension Sayres . . . . $12.00 to 16 00

MEASUREMENTS REQUIRED.

1) Circumference of Thigh.
2)   "   " Knee.
3)   "   " Leg below calf.
4) Length from Knee to below calf.
5)   "   " Knee to Thigh.
6) Right or Left,

1578 Apparatus for Lateral Curvature of the Spine

                         $15.00 to 20 00

Apparatus for Lateral curvature of the Spine.
With elastic Extension Crutches. Price 25.00 to 30 00

One of the main features of this apparatus is the elastic body band, excercising tension against the curvature,

MEASUREMETS, REQUIRED,

1) Sex of patient.
2) Place a thin and flexible strip of lead along the Spine moulding it exactly to the Spine from the seventh Cervical Vertebra to the middle of the Sacrum. With this pattern trace the shape of the Spine on a sheet of paper, marking the points opposite the upper and lower borders of the Scapulae and the crest of Ilium. also the effected parts of the Spine.
3) Length from crest of Ilium to Axilla-right side.
4) Length from crest of Ilium to Axilla-left side.
5) Distance from the center of one Scapulae to the centre of other.
6) Circumference of Body under the Axillae.
7) Circumference of Body between the crest of Ilium and Trochanter Major.
8) Mention whether the convexity of the curve is to the right or to the left.
9) Diameter of the Back from Axilla to Axilla.

Spine Brace Taylors.

Dr. Chas. F. Taylor's Apparatus for Pott's Disease, or posterior Curvature of the Spine.

Price..............................$15.00 to 25 00

Spine Brace Washburns for same purpose

$12.00 to 16 00

Talipes Equinus Apparatus.

Stirrup over toes connected by elastic strap to garter band, Price.................$5.00 to 8 00

For measurements required see Ankle Braces.

Talipes Calcaneus Apparatus.

Stirrup and heel of shoe connected by adjustable elastic band to back of garter band.

Price ....$5.00 to 8 00

Torticollis Apparatus, Davis'.

When ordering this apparatus the shape of the steel head bar should be ascertained, by giving its desired shape to a piece of pliable wire and making a tracing of the latter on a piece of paper.
Price..............................16 00 to 25 00

MEASUREMENTS REQUIRED.

1. Circumference of the Head, passing under the Chin and behind the Ears to the top of the Head.
2. Distance between the Axillae.
3. Distance around the Shoulder, avoiding the Armpits.
4. Circumference of Body between the crest of Ilium and Trochanter Major.
5. Distance from the Navel to the top cavity of the Sternum.

Sayre's Wire Cuirass, according to size 16.00 to 30 00
Same, with Jury Mast....................add 3 00

MEASUREMENTS REQUIRED.

Place patient undressed upon a sheet of paper and with a pencil carefully trace entire outline.

1. Sex of patient.
2. Distance between base of Neck from one side to the other, passing over the Ears and Head.
3. Circumference of Head at Eyes.
4. Circumference of Neck.
5. Length from top of Skull to the Vertebra prominens.
6. Circumference of Body under Axillae.
7. Circumference of Body at the waist.
8. Circumference of Body between crest of Ilium and Trochanter Major.
9. Circumference of Body at Nates.
10. Length from Axillae to Perineum (back),
11. Length from Vertebra Prominens to Perineum.
12. Length from Perineum to Sole of Foot inside
13. Length from crest of Ilium to Sole
14. Circumference of Thigh at Perineum.
15. Circumference of Thigh midway beween Perineum and Knee Joint.
16. Circumference of Calf.
17. Circumference of Ankle.

115

www.ingramcontent.com/pod-product-compliance
Lightning Source LLC
Chambersburg PA
CBHW021942190326
41519CB00009B/1102